WHAT BIOLOGICAL FUNCTIONS ARE
AND WHY THEY MATTER

The biological functions debate is a perennial topic in the philosophy of science. In the first full-length account of the nature and importance of biological functions for many years, Justin Garson presents an innovative new theory, the "generalized selected effects theory of function," which seamlessly integrates evolutionary and developmental perspectives on biological functions. He develops the implications of the theory for contemporary debates in the philosophy of mind, the philosophy of medicine and psychiatry, the philosophy of biology, and biology itself, addressing issues ranging from the nature of mental representation to our understanding of the function of the human genome. Clear, jargon-free, and engagingly written, with accessible examples and explanatory diagrams to illustrate the discussion, his book will be highly valuable for readers across philosophical and scientific disciplines.

JUSTIN GARSON is Associate Professor of Philosophy at Hunter College of the City University of New York. He is the author of *The Biological Mind: A Philosophical Introduction* (2015) and *A Critical Overview of Biological Functions* (2016).

WHAT BIOLOGICAL FUNCTIONS ARE AND WHY THEY MATTER

JUSTIN GARSON

Hunter College, City University of New York

CAMBRIDGE
UNIVERSITY PRESS

CAMBRIDGE
UNIVERSITY PRESS

University Printing House, Cambridge CB2 8BS, United Kingdom

One Liberty Plaza, 20th Floor, New York, NY 10006, USA

477 Williamstown Road, Port Melbourne, VIC 3207, Australia

314-321, 3rd Floor, Plot 3, Splendor Forum, Jasola District Centre, New Delhi - 110025, India

79 Anson Road, #06-04/06, Singapore 079906

Cambridge University Press is part of the University of Cambridge.

It furthers the University's mission by disseminating knowledge in the pursuit of education, learning and research at the highest international levels of excellence.

www.cambridge.org
Information on this title: www.cambridge.org/9781108460026
DOI: 10.1017/9781108560764

© Justin Garson 2019

First published 2019
First paperback edition 2021

A catalogue record for this publication is available from the British Library

Library of Congress Cataloging in Publication data
NAMES: Garson, Justin, author.
TITLE: What biological functions are and why they matter / Justin Garson, Hunter College, City University of New York.
DESCRIPTION: Cambridge, United Kingdom ; New York : Cambridge University Press, 2019. | Includes bibliographical references and index.
IDENTIFIERS: LCCN 2018040445 | ISBN 9781108472593 (hardback : alk. paper) | ISBN 9781108460026 (paperback : alk. paper)
SUBJECTS: LCSH: Biology–Philosophy. | Phenomenological biology.
CLASSIFICATION: LCC QH331 .G28 2019 | DDC 570.1–DC23
LC record available at https://lccn.loc.gov/2018040445

ISBN 978-1-108-47259-3 Hardback
ISBN 978-1-108-46002-6 Paperback

For Elias:
Creative, inquisitive, and funny.
I love you.

Contents

Figures

Acknowledgments

First and foremost, I owe a great debt of gratitude to Karen Neander for several valuable conversations about the topics in this book. Many of my thoughts on function and related topics arose as responses to her work, as will become evident shortly. There are many others to whom I'm especially indebted for discussions that helped to shape the ideas expressed here. These include Carl Craver, Lindley Darden, Dan McShea, David Papineau, Gualtiero Piccinini, and Sahotra Sarkar.

Many people gave me valuable feedback on parts of this book, or on articles that fed into this book. In addition to those listed above, these include Reid Blackman, Robyn Bluhm, Paul Sheldon Davies, Stuart Glennan, Paul Griffiths, Ginger Hoffman, Fabian Hundertmark, Phyllis Illari, Stefan Linquist, Alan Love, John Matthewson, Lenny Moss, Matteo Mossio, Bence Nanay, Anya Plutynski, Gerhard Schlosser, Peter Schulte, Armin Schulz, and Thomas Teufel. I'm also grateful to two anonymous referees who helped make this a much better book. I also wish to thank Hilary Gaskin at Cambridge University Press for taking on this project, as well as Marianne Nield and Sophie Taylor for their valuable editorial assistance.

Finally, my largest debt of gratitude goes to my wife, Rita, for her love, friendship, and support while I wrote this book. I'm also very grateful to my two sons, Elias and Noah, for giving me such an amazing life.

Introduction

This book is about what biological functions are and why they matter. I'll start off in reverse, by saying why they matter.

Functions matter because they're entangled in a dizzyingly wide array of discussions in philosophy and science. Functions lay at the *root*, as it were, of a great number of debates. For example, someone's ideas about what functions are have broad ramifications for thinking about the nature of health, disease, mental illness, mental representation, and mechanisms. Their stance on function can shape their understanding of how different subdisciplines of biology hang together, how to explain diseases, and how to study the human genome. Functions are folded into philosophical debates about biological information, biological trait classification, and even the nature of individuality. Without thinking philosophically about functions, it's almost impossible to think philosophically about nearly anything else in the biological world.

I'll walk through a handful of debates in philosophy and science to show how functions work their way in. Consider the philosophy of psychiatry. Specifically, consider the *basic* problem of philosophy of psychiatry: What *are* mental disorders? Related to that, when psychiatrists decide that a certain condition is a mental disorder, such as major depressive disorder or alcohol use disorder, or gender dysphoria, are they just expressing their values about those alleged conditions? Are they saying, for example, that alcohol use disorder is bad and we wish we could stop it? Or are they simply stating a value-free fact about it, the way you might state that, say, Merriam's kangaroo rat belongs to the genus *Dipodomys*? Or both?

One popular stance (of the naturalist variety) is that what makes something a mental disorder is that it stems from a *biological dysfunction* inside a person. If a man murders another because he's convinced his victim is a disguised alien, he probably has a mental disorder. Something in his brain isn't working *as it should*; there's a dysfunction in his thought processes. But that stance just invites further questions: What are functions? What are dysfunctions? And who gets to decide whose brain is functioning well and whose is functioning poorly? In my view, it's impossible to make progress on the nature of mental illness before getting clear on what functions are. (In Chapter 11, I'll come back to mental disorders.)

Turn to the philosophy of mind. Consider, specifically, the problem of *intentionality* or *aboutness*. What is it for a thought, which is inside my skull, to be about something outside my skull? What's the nature of this invisible relationship – if it's a relationship at all – that connects the two things? Starting in the 1980s, a group of philosophers began advocating for an evolutionary, naturalistic approach to intentionality. According to this view, that thing that gives me the power to have the thought, "Kampala is the capital of Uganda," is, at its most primordial, the same thing that gives a toad the power to register the presence of edible worms in its visual field. This viewpoint, known as *teleosemantics*, holds that mental representation is ultimately grounded in biological functions, which are, in turn, grounded in natural selection.

One of the real virtues of teleosemantics is that it shows how organisms can misrepresent their environments. When an organism misrepresents something – say, a bird misrepresents a crocodile as a log – there's a device in its brain that fails to perform its biological function; it's not responding *as it should* to predators. My own view is that teleosemantics is right but that many of its proponents go awry because they cling to an overly narrow conception of what functions are. This conception, *the traditional selected effects theory*, holds that a trait's function is just whatever it was selected for by natural selection (or a related selection process). The traditional selected effects theory generates spurious problems for teleosemantics, problems that have led some philosophers to turn their backs on it entirely. Once we clarify what functions are, some of these problems simply go away (see Chapter 12).

The functions debate doesn't just matter for philosophy. It matters for biology, too. It has deep implications for how we study the human genome. A few years ago, an unusually heated debate broke out between a group of geneticists. They began arguing about what proportion of the human genome is *functional*. One side of the debate, represented by the ENCODE Project Consortium, maintained that around 80 percent of the genome is functional. The other side of the debate, represented by the traditional "junk DNA" theorists, insisted that only 5–15 percent is functional. In response to ENCODE's (probably exaggerated) claims, the junk DNA scientists argued that ENCODE was ramming together two very different senses of "function," the *selected effects* sense and the *causal role* sense. They began citing the work of philosophers of biology, including my own, to support their contention. Thinking deeply about functions has immediate payoffs for the inner workings of biology itself, and not just for philosophers who like thinking about it. (And I mean "payoff" in the most literal sense; proponents of ENCODE claim that mapping these genomic functions will lead to major medical breakthroughs, a notion that plays heavily into the mechanics of funding. Claims about genomic function have a literal cash value.) I'll return to this project in Chapter 9.

I've indicated why functions matter, and I'll return to these debates in the book; now I'll say what they are.

This book sets out a novel theory about biological functions, the *generalized selected effects* theory (GSE). As the name suggests, it's related by descent to the traditional selected effects theory but drops some unnecessary limitations on the latter and draws out a principle that was buried inside it. The book also works out the implications of GSE for the problems I mentioned above. In particular, it recommends a novel way of thinking about mechanisms, mental disorders, and intentionality. It also reframes the debate about function pluralism, a topic that plays heavily into the ENCODE controversy. One of my main goals in this book – aside from convincing you that my theory is right – is to inject new life into the functions debate and show why it's so essential for thinking about other big problems in philosophy and science.

What exactly is GSE? GSE states that the function of a trait is whatever it did, in the past, that contributed to the trait's *differential reproduction or differential retention within a population*. It's an unabashedly historical account, since it claims that nothing in biology can have a function until it's gone through a few rounds of selection. GSE merges three key principles: differential reproduction, differential retention, and population.

The first part, differential reproduction, captures the core insight of the traditional selected effects theory – namely, that a trait can acquire a function because of how it caused the trait to multiply. The function of zebra stripes is to deter biting flies, since that's what helped the striped zebras out-reproduce the stripeless ones. The second part, differential retention, leads us out of the evolutionary realm and into the realm of development. Synapses in the brain, for example, don't reproduce. For them, success means persisting better than your neighbor. Hence the "generalized" part of the generalized selected effects theory: It includes everything the traditional selected effects theory does, and many other things in addition. It shows how antibodies can acquire new functions through antibody selection, how synapses can acquire new functions through synapse selection, and how behaviors and behavior dispositions can acquire new functions through operant conditioning.

Others have gestured toward the possibility of generalizing the selected effects theory to include development, but they were hindered in this task because they tried to fit all functions into the mold of differential reproduction. Millikan, for example, says that trial-and-error learning creates new functions because it helps behaviors reproduce over one's lifetime. That's nonsense any way you read it. Generalizing the theory requires more than a gentle tweak; it requires reconfiguring it from the ground up and showing why, on the basis of first principles, that reconfiguration is correct.

The third part, *in a population*, simply teases out a principle that's dormant in the very idea of selection. For selection always takes place within something like a population: that is, a group of individuals that impact each other's chances of survival, helping or hurting each other, as the case may be. The reason it's worth making that *implicit* commitment *explicit* is that it solves, at a stroke, various complaints people have lodged against the traditional selected effects theory as well as my own. Some have argued that the theory forces us to give functions to all sorts of things that don't deserve them, such as clay crystals, ball bearings, or piles of rocks, but since a pile of rocks isn't a population, its parts don't have functions.

The book is broken into three main parts: background, theory, applications. The first part consists of Chapters 1 through 3 ("Background"). These set out the foundation for the theory, and show why functions must be selected effects. In Chapter 1, I consider a puzzling feature of ordinary biological usage: namely, function's *explanatory depth* – sometimes function statements are causal explanations for traits. When biologists say that the function of the zebra stripes is to deter biting flies, for example, they're trying to explain *why zebras have stripes*. In Chapter 2, I argue that *if* functions are selected effects, they have explanatory depth. In Chapter 3, I argue the converse: namely, if functions have explanatory depth, then they're selected effects. If we take explanatory depth seriously, then the traditional selected effects theory, or something in its neighborhood, has no equal. I also defend the theory from a host of objections.

The second part of the book ("Theory"), Chapters 4 through 8, sets out GSE and explains why it's preferable to the traditional selected effects theory. It defends the generalized selected effects theory from various objections, it solves the problem of function indeterminacy, and it explains what dysfunctions are. In Chapter 4, I review how philosophers have tried to extend the traditional selected effects theory to other phenomena, such as learning, and I offer my diagnosis of where they went wrong – namely, they tried to fit all functions into the framework of reproduction or copying. Chapter 5 focuses on synapse selection, since it's a plausible case where functions come about simply by virtue of *differential retention* – one synapse outlasting another – even when there's nothing like reproduction happening. Chapter 6 presents, at long last, the generalized theory and defends it from seven objections, presented in order of increasing difficulty. Chapter 7 raises the problem of function indeterminacy and shows why solving it matters for biomedicine and teleosemantics. It defends the idea that proper functions are *proximal* functions. Chapter 8 says what dysfunctions are, and shows why GSE is preferable to other views on this score, such as Boorse's biostatistical theory of function.

The third part of the book ("Applications"), Chapters 9 through 12, applies GSE to problems in philosophy of biology, philosophy of medicine and psychiatry, and philosophy of mind. This is where the theory pays off in practical ways. Chapter 9 deals with the problem of function pluralism, which plays into the ENCODE debate, and says why the received version of pluralism, *between-discipline pluralism*, is wrong. Between-discipline pluralism tries to fit all biological uses of "function" into two categories, the *selected effects* sense and the *causal role* sense, and it tries to divide up biological disciplines, like genetics, neuroscience, and ethology, into two groups depending on which sense of "function" is more prominent. Chapter 10 delves into the topic of mechanisms and mechanistic explanation. It shows how mechanisms have a hidden, functional side, and once we draw out this functional aspect of mechanisms, we can make sense of how mechanisms break down. It also lays out a program for biomedicine: As a rule, don't look for mechanisms for diseases; instead, show how diseases come from breakdowns in mechanisms for functions. Chapter 11 draws out the implications of GSE for the philosophy of psychiatry and shows how one popular theory of mental disorder, the harmful dysfunction analysis, is wrong. If GSE is right, then many of the mental disorders that plague us, like generalized anxiety disorder, addiction, and even the delusions of schizophrenia, might not involve dysfunctions; maybe they're functional in their own right. Chapter 12 works out the consequences of GSE for teleosemantics, and argues that, if we accept GSE, we can solve a long-standing problem, the *problem of novel contents*. The chapter also defends a solution to a related problem, the problem of distal content.

This book is limited in one major way. It's not intended as a systematic survey of the vast literature surrounding biological functions, although I'll introduce many alternative theories along the way. I've already written that survey, *A Critical Overview of Biological Functions* (Garson, 2016), and found no need to do so again. For someone who wants an exhaustive catalog of various ideas people have entertained about functions for the last eighty years or so, that is the place to go, but for readers who want to know what functions are, and why they matter to other philosophical issues, this is the place to turn. The other can be seen as a helpful companion volume to this one.

I've drawn from numerous published, or forthcoming, papers and books. This work is indicated in the references section as Garson (2010; 2011; 2012; 2013; 2015; 2016; 2017a; 2017b; forthcoming a; forthcoming b; forthcoming c) and Garson and Piccinini (2014). But as far as the actual writing goes, this book was written entirely "from scratch." Its real virtue is that it has allowed me to distil the essential ideas of previously published work, expand them in certain ways, contract them in others, and put them together into a simple and, I hope, attractive whole.

PART I

Background

CHAPTER I

The Strangeness of Functions

Why do zebras have stripes? Biologists have argued about this since at least Darwin's time. Darwin himself dismissed the popular view that the stripes' purpose is camouflage: "The zebra is conspicuously striped, and stripes on the open plains of South Africa cannot afford any protection" (1871, 302). Others insist that stripes aren't there for camouflage but for cooling the animal (Larison et al. 2015). They think the black-and-white pattern chills the air around it. A third idea is that stripes play a role in social cohesion; the striped pattern draws zebras together into herds (Macdonald 2009, 689). A fourth possibility is that zebra stripes have no function at all (although I don't know of anyone who argues this in the literature). Maybe they're as biologically pointless as birthmarks, freckles, and chin clefts.

Recently, an American biologist, Tim Caro, threw his weight behind a newer idea (Caro et al. 2014; but see Harris 1930). He thinks the stripes' purpose is to deter biting flies. One particular family, the glossinids (commonly known as tsetse flies), is particularly troublesome, since it harbors a parasite responsible for African trypanosomiasis – the infamous sleeping sickness. Field and laboratory studies suggest that tsetse flies and other biting flies are averse to striped surfaces. Perhaps zebras use stripes to exploit this neurological quirk of the tsetse fly. Caro's hypothesis about the stripes' function is based on a mix of historical, geographical, and laboratory evidence, although the whole subject remains mired in controversy.

The parts and processes of the tsetse fly have functions, too. The tsetse fly is a family of bloodsucking flies that inhabit Central Africa, from the Sahara in the north to the Kalahari in the south. Unlike ordinary houseflies, it has a long, hollow proboscis. The tip of the proboscis is lined with tiny, sharp teeth, like a knife's serrated edge (see Krenn and Aspöck 2012, 111, Figure 8). The fly repeatedly prods an animal's thick hide until it draws blood. Its pharynx functions as a pump that sucks up the nutritious broth. A second pump shoots saliva into the wound in order to stop the blood from coagulating. The trypanosome parasite, *T. brucei*, lives in the saliva.

The parasite *T. brucei* has functions of its very own. It is one of many unicellular species of African trypanosome, and it resembles a tiny seahorse. It is the parasite responsible for sleeping sickness. Its coat contains millions of proteins called *variant surface glycoproteins*. The function of these proteins is to help *T. brucei* evade detection by the host's immune system (Horn and McCulloch 2010). The coat's genetic makeup is constantly changing: By the time the host's immune system learns to recognize one coat, *T. brucei* has morphed into another. As one geneticist described it to me, it is like changing hats, and the parasite changes its hat about once a week.

Functions are ubiquitous in the living world. Sometimes they harmonize; sometimes they clash. What are functions? At first glance, functions seem easy to understand. If functions are easy to understand, we should be able to give a clear and satisfying account of what they are. Instead, we find puzzles, and even contradictions, that drive us deeper into the nature of the living world.

When I ask biologists what functions are, I often get a similar response: "A trait's function is just what it does." Sometimes these biologists seem perplexed, and even mildly annoyed, to be asked a question like that. Hearts pump blood. That is what they do, so that is their function. Zebra stripes deter flies. That is what *they* do, so that is their function. The tsetse flies' labellar teeth puncture skin; *T. brucei's* glycoprotein coat tricks the host's immune system. Functions are simply doings.

Sadly, the biologists' simple account can't be right – for two reasons. First, traits do many things that aren't their functions. Noses help us breathe; they also hold up glasses, but their function is to help us breathe, not hold up glasses. Holding up glasses is a lucky benefit, or side effect, but not a function. Zebra stripes entertain safari guests, but that's also not their function. To use philosophical lingo, the fact that stripes entertain safarigoers is an "accident" and not a function. A good account of function should help us understand how functions and accidents differ.

Here's a second problem with the simple account that says a trait's function is just what it does. A particular instance of a trait – my stomach, your heart – can have the function of doing something even if it can't actually do that thing. If my stomach shuts down because of a drug overdose, it can't digest food. Yet it has the function of digesting food (it's a stomach, after all); thus it has a function it can't perform. It's "dysfunctional" or, if you prefer, "malfunctioning." Philosophers sometimes call this feature of functions – that it's possible for a trait to have a function it cannot perform, that it can dysfunction or malfunction – the "normativity" of functions. A good theory of function should make sense of this normativity, too.

We have to be careful here. Scientists use the word "function" in different ways; one theory won't fit all uses. I need to home in on the sense I'm after. In one sense of "function," functions are just *effects*. Climate change is a function of deforestation. Poor academic performance is a function of malnutrition. That isn't the sense of "function" I want to know about, and it's not the one that's prominent in biology. That sense of "function" doesn't let us distinguish functions and accidents; nor does it have a normative side. Whenever I use the term "function" without qualification, I mean it in the ordinary biological sense, where functions differ from accidents and where things sometimes malfunction. In Chapter 9, I'll return to the problem of how different senses of "function" fit together.

Here's the plan for this chapter. In Section 1.1, I'll turn to ordinary biological usage to extract a vital clue about functions: namely, functions have *explanatory depth*. When biologists give functions to traits, they often purport to give causal explanations for why those traits exist. By meditating on this one feature of functions, we can solve the other main puzzles of function. The problem is that, at first glance, it's hard to see how functions can actually explain anything; this is the so-called problem of *backwards causation* (Section 1.2). Section 1.3 will survey some of the more adventurous ways people have tried to solve backwards causation. It will also introduce the idea that, to solve backwards causation, functions should be selected effects. In Section 1.4, I'll be a bit more precise about what explanatory depth amounts to: that is, what functions are *supposed* to explain, whether or not they actually do so. In Section 1.5, I'll set out the ground rules of the game: What exactly is a theory of function, and what kinds of evidence should one draw on to support such a theory?

1.1 Functions and Explanations

What are functions? One clue comes from considering their role in explanations, in the practice of asking and responding to why-questions in science and in everyday life. Sometimes, when people say that a trait has a function, they're trying to explain why that trait exists. For example, sometimes, when scientists wonder about the function of zebra stripes, they're just wondering why zebras have stripes (rather than, say, being monocolored like some horses). And sometimes, when scientists argue with each another about the stripes' function, they're just arguing about how zebras came to have stripes. Maybe if we think about how functions fit into explanations, we'll come closer to understanding these other puzzling features of function. (In Section 1.5, I'll defend my method of figuring out what functions are: Examine how biologists use the term "function," and then step back and figure out what functions must actually be to support their usage.)

It's helpful to have some concrete examples in front of us. Tim Caro and his colleagues (2014), whom I alluded to earlier, wrote a paper simply entitled "The function of zebra stripes." It appeared in a major scientific journal, *Nature Communications*. From the outset, Caro makes it clear that solving the riddle of the stripes' function *just amounts to* explaining why zebras have stripes. Three pieces of textual evidence back up this interpretation. First, Caro writes of five different "functional hypotheses" about zebra stripes; he also calls these "factors proposed for driving the evolution of zebras' extraordinary coat coloration" (2). For Caro, offering a "functional hypothesis" about stripes and making a conjecture about why stripes evolved are one and the same thing. It's not that there are two questions, one about functions and one about origins. There's one question that can be posed using different words.

Second, to support his hypothesis about the stripes' function, Caro collected historical data about the distribution of tsetse flies in the regions he studied, as well as the historical distribution of various predators. Given that he wanted to show why stripes evolved, this was a sensible thing to do. Tsetse flies must have actually lived in the zebra's habitat when stripes evolved; otherwise, his "functional hypothesis" would be demolished. If Caro didn't care about why stripes evolved, why bother collecting historical data, which can be quite time-consuming and labor-intensive?

One final fact about Caro's paper seems noteworthy. A few years later, he and his colleagues wrote another paper with the elegant title, "Why is the giant panda black and white?" This one appeared in the journal *Behavioral Ecology*. It's plausible to think that the question he was asking about zebra stripes is the exact same kind of question he was asking about the panda's unique constellation of markings. The difference is verbal: One title is phrased in terms of functions, and the other is phrased in terms of why a trait exists. My point isn't that Caro is right about the function of stripes. My point is that, in some cases, function statements purport to be explanations. They have explanatory depth.

Here's another example of how biologists fuse functions and explanations. In his book, *Neural Activity and the Growth of the Brain* (1994), the neuroscientist Dale Purves discusses how hard it is to decipher the function of animal markings. I'll cite a long passage because it reveals the explanatory depth of functions:

> Skin and fur markings are so striking that it is natural to assume that they must reflect some fundamental *function* of the integument. The major *purposes* of the skin, however, are temperature control, water regulation, and protection from infection. In fact, zoologists have often found it rather hard to decipher the *role* of particular animal markings. Such patterns are sometimes used for camouflage or sexual attraction, but more often than not it is difficult to say just *why they are there*... (30; my emphases)

What is so striking about the passage is that Purves uses the expressions, "purpose," "function," "role," and "why they are there," interchangeably. In his usage, to state something's function just amounts to saying why it's there. Purves doesn't go so far as to say that all markings have functions. Maybe some markings, like birthmarks or freckles, are purposeless. My point is that, sometimes, when scientists give functions to traits, they're trying to explain why those traits exist.

Here's a final example of how biologists tie functions to explanations. The Harvard molecular biologist Sean Eddy, in a discussion of the concept of "nonfunctional DNA," tells us that "by nonfunctional, we mean 'having little or no selective advantage for the organism'" (2012, R898). This makes it sound as if the function of a stretch of DNA is simply whatever it does that promotes the organism's relative fitness, regardless of how it got there. Later, however, he clarifies that when we argue about whether stretches of DNA are functional, we're arguing about "whether *they're there* primarily because they're useful for the organism (R898, emphasis mine)." So, for Eddy, when we say something's functional, we're not saying (or not only saying) that it does something or other to help us out, but that the fact that it helps us out in the specified way is the reason it exists.

Scientists aren't alone in using "function" with explanatory depth. Laypeople do, too. Several newspapers reported Caro's work on the function of stripes, and they used expressions like "why zebras have stripes," or "why stripes evolved," synonymously with "the function of stripes." Everyone seemed to understand that Caro's paper, "The function of zebra stripes," was about why zebras have stripes. The idea that functions are explanations is not some philosopher's invention. It's a robust feature of how scientists and laypeople alike think and talk about them.

This is a remarkable feature of functions, and one we shouldn't brush under the rug. In fact, philosophers of science have puzzled over the explanatory depth of functions ever since they started thinking seriously about them. I trace the modern functions debate back to the philosophers Carl Hempel (1965) and Ernest Nagel (1953). They agreed that function statements often *purport* to be explanations; they took this explanatory ambition as a plain fact of ordinary biological talk. They disagreed, however, about whether functions *actually* explain anything: that is, whether this explanatory ambition is ever fulfilled. Before we get tied up in debates, though, we should nail down exactly what is at issue. If function statements are supposed to be explanations, what is supposed to explain what? And what kind of explanation is on offer?

Consider how the popular press reported Tim Caro's work. According to the journalists, Caro and his colleagues discovered why zebras have stripes. The reason zebras have stripes is that stripes deter biting flies. It appears that, in Caro's functional hypothesis, the fact that stripes deter biting flies is supposed to explain why zebras have stripes.

To use philosophical lingo again, an explanation splits into two halves: the *explanandum* and the *explanans*. The explanandum is the fact or event one wishes to explain: why zebras have stripes; why Booth shot Lincoln; why rocks fall to the ground when you let them go. And the explanans is the fact or event that does the explaining: stripes deter flies; Booth wanted to extend the war; space-time is curved. If the statement, "The function of zebra stripes is to deter biting flies," is an explanation, the explanans is: Stripes deter flies. The explanandum is: Zebras have stripes.

> Explanans: Stripes deter flies
>   explains
> Explanandum: Zebras have stripes

Assuming that functions are explanatory, what kind of explanation is on offer? Explanations fall into different categories. There are, for instance, mathematical explanations, causal explanations, reductionist explanations, and statistical explanations (see Salmon 1989). The most natural answer here – the one that best fits the examples above – is that function statements are *causal* explanations, since they have to do with how things came to be. In a typical causal explanation, one explains why one event happens by pointing to an event that came before it. Why did Selena make the dean's list? Because she worked very hard in school. Why is there a coffee stain on the carpet? Because I knocked the mug over when I was reaching for the corn flakes. (Later, I'll consider, and reject, the idea that functions are explanatory in some noncausal sense.)

If function statements are explanations of the causal sort, then when we say the function of the zebra's stripes is to deter biting flies, we're saying that the fact that stripes deter flies causes zebras to have stripes. At any rate, that's the most natural way of interpreting biologists like Caro, Purves, and Eddy:

> Explanans: Stripes deter flies
> *causes*
> Explanandum: Zebras have stripes

From this perspective, when scientists give functions to traits, they're just giving extremely compact causal explanations for why those traits exist. Any theory of function that takes the causal-explanatory role of functions seriously is called an *etiological* approach to function, because "etiology" refers to the study of causes. The etiological approach to function isn't a single theory. It's a family of theories joined by the idea that function statements are causal explanations for the existence of traits. Its *locus classicus* is Wright (1973) – although as we'll see, Wright's specific version was somewhat off the mark.

This way of thinking about functions, where functions are just condensed causal explanations, is very attractive. In addition to reflecting the way biologists think and talk, it illuminates the two puzzling features of function we started with. (I'll come back to this point in the next chapter.) First, it shows how functions differ from accidents. The reason a function of the nose is to help us breathe, and not hold up glasses, is because the fact that noses help us breathe explains why we have them. The fact that noses are good at holding up glasses isn't why noses are there. The etiological approach also makes sense of function's "normativity" – that is, the possibility of malfunction or dysfunction. Something dysfunctions when it cannot do the thing that explains why it's there.

As an aside, when I say functions are "normative," I don't mean anything very nuanced or sophisticated. All I mean is that it's possible for something to dysfunction. Once we've explained how it's possible for something to dysfunction (see Chapter 8), we've explained function's "normativity." In my usage, there's nothing else hiding behind the word, nothing having to do with values or goals, oughts and shoulds, prescriptions or commands, the good or the just. Sometimes when I explain to people how dysfunctions are possible, they say things like, "Yes, I see, but how do you explain function's *normativity?*" In my preferred usage, there is no additional question here, but I don't wish to legislate usage for everyone else.

1.2 Backwards Causation

Despite its merits, etiological approaches to function seem to suffer a major drawback. This is known as the problem of "backwards causation" (e.g., Ruse 1973, 176). On the face of it, the fact that stripes deter flies cannot possibly explain why zebras have stripes. In a standard causal explanation, the relationship between cause and effect is a before-and-after one. In order for my knocking over the coffee cup to cause a carpet stain, I first knock over the cup, and then the stain appears, and not the other way around. (There might be exceptions to this rule. Kant noted that when a ball sits on a cushion and causes an impression on it, the ball and the cushion exist simultaneously. I won't linger on this because it isn't relevant to the typical biological examples.)

In order for the fact that stripes deter flies to cause zebras to have stripes, the two facts would then have to stand in a before-and-after relationship. It would have to be the case that, first, zebra stripes deter flies and, second, zebras have stripes, but that's the reverse of what happens in the real world. For a zebra to use its stripes to deter flies, it must already have the stripes. Hence the "backwards" part of the problem of backwards causation. In order for the fact that stripes deter flies to cause zebras to have stripes, there would have to be a peculiar causal relationship in which later events (a zebra using its stripes to deter flies) cause earlier ones (zebras have stripes). The etiological approach to function seems to flip the normal chronological order of causation. It violates a cornerstone of our understanding of cause and effect, at least as it applies to things outside of the weird quantum realm.

Here's a simple, no-nonsense solution to backwards causation: Flatly deny that function statements are causal explanations. This is what Hempel (1965), in effect, did. In fact, he denied that they're explanations at all. As he put it, "the information typically provided by a functional analysis of an item *i* affords neither deductively nor inductively adequate grounds for expecting *i* rather than one of its alternatives" (p. 313). In short, function statements don't actually explain why the functional item exists. Hempel admitted that scientists often think that functions are explanatory. He just thought they were in the grip of a cognitive illusion: "The impression that a functional analysis…explains the occurrence of *i*, is no doubt at least partly due to the benefit of hindsight: when we seek to explain an item *i*, we presumably know already that *i* has occurred" (ibid.).

Many philosophers have chosen to follow Hempel in denying that functions statements are causal explanations. They have worked out a vast array of *nonetiological* accounts of function instead. Fitness-contribution theories, such as the propensity theory and the biostatistical theory, hold that a trait's function has to do with its present-day contribution to fitness. Causal role theories say it has to do with the contribution a part makes, in tandem with other parts, to an interesting system capacity. Modal theories say the function of a trait has to do with its behavior on nearby possible worlds. Some versions of the organizational theory hold that a trait's function has to do with how it contributes to the persistence of that very trait. (Most of these will come in for scrutiny at various points of the book.) I have mapped out those positions in detail elsewhere, and outlined their relative strengths and weaknesses (Garson 2016). The thread that joins them together is their shared rejection of the idea that functions are causal explanations for traits.

I think it would be unfortunate if functions never provided successful causal explanations for traits, since so many scientists, science writers, and journalists, think they do. How could they all be so wrong? The fact that it would be unfortunate, however, shouldn't count for much. Sometimes, philosophy reveals that our deeply held assumptions about language, ethics, or reality, are mistaken; maybe this is just one of those occasions. Let's not give up too quickly though. We can at least try to find a coherent theory of function that shows how functions can be successful causal explanations. If our best efforts in this direction repeatedly end in frustration, we should be ready to embrace a *nonetiological* theory instead.

1.3 Theism and Fictionalism

Before throwing out the etiological approach because of backwards causation, here's a thought that should give us pause. At least sometimes, the effect of an item does explain "why it is there" – that is, why it is where it is. Suppose I ingest some beneficial bacteria (such as *Lactobacillus*) to promote good digestion, because I know it will raise gut acidity. In that case, the *Lactobacillus* is there, in my gut, *because* it promotes acidity. There's no mystical or mysterious backwards causation taking place. The fact that bacteria are good at promoting acidity caused me to put them in my gut.

Consider a somewhat more extreme case. Suppose we rescue a species from extinction because of some benefit it provides. In the early 1990s, southern California zoos decided to rescue California condors (*Gymnogyps californianus*) from extinction because they are huge, magnificent creatures, and they are good scavengers, too. The reason condors exist – exist *now*, *today* – is because they are huge, magnificent, and good at scavenging. If they possessed none of those features, we would no longer enjoy their presence. There's no mysterious backwards causation taking place when we say that condors exist because they are huge, magnificent scavengers.

Unfortunately, this solution to backwards causation doesn't apply to the ordinary biological cases, like the zebra's stripes, the tsetse fly's labellar teeth, or the ever-morphing coat of *T. brucei*. We solved backwards causation by bringing in intelligence and mind – that is, the intelligence of human beings and their mental states like beliefs and desires. The fact that *Lactobacillus* promotes gut acidity caused my *belief* that it promotes gut acidity; my *belief*, combined with my *desire* for a healthy digestive system, caused me to put them in my stomach. At best, this solution only applies to intelligent beings, and the things we do or make.

Of course, if we were willing to embrace theism, the problem of backwards causation would disappear. Perhaps there is a God who freely shapes the biological world, just as I freely shape my gut biota. Maybe God knew that zebra stripes would deter flies, and God gave stripes to the zebra for that reason, out of the kindness that defines God's very nature. Were that the case, we could affirm that zebras have stripes because stripes deter flies. (For that matter, perhaps the world as we know it is a mad scientist's computer simulation, and this scientist gave the zebra stripes just to confuse and bewilder our brightest minds. In that case, the stripes would still have a function, though one very different than we imagined.)

Considerations such as these led some philosophers to an adventurous position. They think that if there are any biological functions in the natural world, an intelligent being must have put them there. In other words, some consider it a conceptual truth that function requires intentional design. One can't accept functions, but reject God, in one and the same breath.

The philosopher Alvin Plantinga (1993, 214) deploys this idea as part of an innovative argument for God's existence (though one that echoes Aquinas, as he freely admits). His first premise is that there are functions in the natural world, as any biologist will tell you. (I invite you to track some down and ask them yourself.) His second premise is that, as a conceptual truth, functions require design. Thus, God is real, and a millennia-old controversy is solved.

Michael Ruse (2002), like Plantinga, thinks function requires design, but he reaches an even more adventurous conclusion. He agrees that, as a point of definition, if anything has a function, it must have been designed. Biologists talk about functions, he says, "because organisms. . .are taken to be design-like: they are taken to be as if they were artefacts, or parts of artefacts, created by conscious intelligences in order to serve certain ends" (2002, 37). Instead of embracing God, he denies functions. Strictly speaking, functions don't exist. When biologists talk about functions, they are trading in metaphors; biologists are playing a game of make-believe (p. 40). However – and this is the strange part – Ruse thinks biologists shouldn't stop playing this game of make-believe. That's because functions have "key heuristic value" (p. 46). When biologists look at organisms as if they were designed, they often discover new things about them. Philosophers often call this sort of view "fictionalism."

I want to avoid both of these extremes, theism and fictionalism. I'd like to think that functions exist, just like biologists think they do. The idea that science should embrace known falsehoods as an engine of discovery runs against its core mandate. I'd also like to think one could acknowledge that functions are real without taking a position on the existence of God. Finally, I want to acknowledge the explanatory depth of functions – that is, function statements are, sometimes, correct causal explanations for traits, as biologists like Caro think they are. What to do?

Fortunately, there is a solution. We can solve the problem of backwards causation by appealing to selection processes, not design. In the next chapter, I'll lay out the core argument for the traditional selected effects theory. This theory, in its most unadorned version, says that a trait's function is what it was selected for. I'll also defend the integrity of that solution from common complaints. The crucial point here is that the reason selected effects theorists tie functions to selection is neither because they think natural selection is the only force of evolutionary change, nor because they think natural selection is itself a source of design, nor because they don't understand the history of biology. (All of these are accusations that philosophers have leveled against selected effects theorists.) Rather, the selected effects theory is the best way to solve a conceptual problem that has plagued the functions debate for over sixty years: How do functions explain anything?

1.4   Being There

A few more aspects of function deserve clarification before moving on. Functions purport to explain why a trait "is there." Can we clarify this "being there"? What exactly are function statements supposed to explain? To be precise, when we say that some activity of a trait is its function, we purport to give a causal explanation. The explanandum – the thing we're trying to explain – can take two forms. Sometimes, we want to explain a fact about some particular entity: namely, why that entity has that trait. (Why does my rat Gemini stand on his hind legs when he wants some celery?) Other times, we want to explain a fact about a population. We want to explain why some, or most, or all members of a population have a trait. (Why do people have noses? Why do zebras have stripes?) Some, like Buller (1998, 522), think function statements are primarily trying to explain facts about individuals (why Gemini stands on his hind legs). Others, like Neander, (1991, 174), think functions statements are primarily trying to explain facts about populations (why people have noses). I don't see any good reason for restricting the explanandum of function statements either way. Sometimes, function statements are about individuals, and sometimes they're about populations.

Another question that comes up, by way of clarifying the explanandum, is this: What is it for an individual or group to "have" a trait? To say that an entity (individual or group) has a trait is to say that the trait is a physical, or psychological, or behavioral feature of that entity. (I'd use the term "part," but that's a bit too restrictive since it connotes physical parts. "Feature" is broad enough to capture behavioral or psychological characteristics, too.) Crucially, for an item to have a function, it must be a feature of a system, like an organism, cell, or group; but the system, considered as a whole, has no function. Hence, we can ask about the function of the zebra's stripes, or the function of the prairie moles' grooming behavior, or whether depression has a function in mammals. We can wonder about the function of group traits, like V-formation in a flock of geese, or predator signaling in a vervet colony. But the prairie mole itself has no function; nor does the zebra; nor does the flock.

This restriction – functions belong to features of systems but not to the system itself – is enshrined in ordinary biological usage, just like the claim that functions have explanatory depth. To see this, one need only consider the following thought experiment. When Caro and his colleagues wrote a paper entitled "The function of zebra stripes," the title was easy to understand. Had they written a paper called "The function of *zebras*," nobody would have known what they meant. Similarly, when he wrote, "Why is the giant panda black and white?," everyone knew what he was getting at. Had he written, instead, "Why are there pandas?," it would not have been published by a major scientific journal. (In the tradition of natural theology, which reached its apex with William Paley's 1802 book by the same title, one could still pose questions in this matter; e.g., why do mosquitoes exist? Biologists stopped asking questions like that by the middle of the nineteenth century for reasons that need not be rehearsed here.)

This way of thinking about functions, where functions belong, first and foremost, to *features* of individuals or groups, has a drawback. It prevents certain things from having functions that we might think of as having them. Consider a beaver dam or a honeycomb (Griffiths 1993, 416). Surely, the honeycomb has a function for the bee population: namely, to contain larva and store honey, but the honeycomb is not a feature of any particular bee or even a feature of a whole population of bees. According to my way of thinking, the honeycomb does not have a function, or at least not in the way that the bee's stinger has a function, but this is an acceptable limitation. We can say everything we'd like to say about functions without giving functions to honeycombs and beaver dams. Instead of talking about the function of the honeycomb, we can talk about the function of the comb-making behavior, which is a feature of the bee. There's no need to insist that honeycombs or beaver dams have biological functions of their very own. (As one reviewer pointed out, we might also solve this by considering the honeycomb as part of the bee's *extended phenotype*, in which case we could talk about the function of the honeycomb just as we talk about the function of the comb-making behavior – see Dawkins 1982. I won't pursue this suggestion here.)

Now, beaver dams and honeycombs do have "functions," after a certain manner. A honeycomb has an *artifact* function, much like the components of my old flip phone, or a car's engine, or clay pottery from the Neolithic era. As I'll discuss in the next chapter, artifact functions aren't biological functions, and we shouldn't force them into the same mold. We need one theory for biological functions, and another theory for artifact functions. It seems to me that beaver dams acquire their functions, in the first place, because individuals design, create, and use them; they don't get their functions in the same way that features of organisms do.

I realize that some entities have both biological and artifact functions, such as an edited segment of DNA or a marijuana-sniffing hound. This complicates the picture, but it doesn't change the fact that biological and artifact functions are different kinds of things. Why do I think this? Because the best theory of function on the market is the selected effects theory (to be precise, a particular version called the generalized selected effects theory) and that theory implies that, as a rule, artifact functions aren't biological functions, as I'll show in the next chapter. I don't have any theory-neutral argument for why biological and artifact functions are distinct. I treat their distinctness as an interesting consequence of the selected effects theory.

1.5   Rules of the Game

Before setting out to discover what functions are, we should lay down some ground rules for how to proceed in this intellectual endeavor. What is a "theory of function" supposed to be? How do we decide between competing theories of function? What kinds of evidence should we take seriously?

Let me go back to the beginning, but this time from a *meta* perspective. I started with a question: What are functions? By about halfway through the book, I will have defended an answer of the form, "functions are X." What is this statement supposed to be? Is it a report about how people think about functions? Or is it a report about what the world is like regardless of what anyone thinks, like "water is H2O"? And if it's a report about how people think about functions, is it a report about how scientists think about them? Or is it about how ordinary people think about them? Or is it something like a recommendation for how we *ought* to think about them, regardless of what the world is like and regardless of what anyone happens to think now? Sometimes when philosophers seem to disagree about what functions are, they're really disagreeing about these meta questions. For example, I take it that when Ruse and Plantinga say that functions are intended effects, they mean to report something, first and foremost, about how ordinary people think about functions, about the concepts they have in their minds, and they might even be right – but that might still be irrelevant to my own project.

Here, I take the second approach: For me, a theory of function is a report about what exists in the world; it's not a report about what's in people's heads. Coming up with a theory of function is just like coming up with a theory about what gold is, or what aluminum is, or what spiders are. To know what aluminum is, you wouldn't, first of all, canvas people's opinions about it. You would get a chunk of it and start poking and prodding it.

Other people might be more interested in understanding our concepts of function, and I applaud their endeavors, but that's not what I'm after.

There's an extra layer of complexity here, however, since you can't entirely separate these two projects – that is, of figuring out what functions are, and figuring out how people think about them. Sadly, functions aren't like aluminum or spiders. You can't just collect a sample and start poking at it. Functions are more abstract; they're less tangible. So you need a somewhat more abstract method to get at them. It seems to me that the very best way to figure out what functions are is to look at how biologists think about them. After all, if anyone knows what functions are, it's the biologists. For all that, I'm not trying to come up with a theory of what biologists think; I'm turning to biological usage to extract vital clues about what functions are. Looking at how biologists think about functions is a means to an end, not the end itself.

Now, on to the question of evidence. What kinds of evidence would prove that I'm right? There are three kinds of evidence I take seriously, ranked in terms of how seriously I take them: ordinary biological usage, ordinary biological practice, and bald intuitions. First and foremost, I think the best way to approach functions is to look at ordinary biological usage – that is, how biologists talk about them – as that talk is captured in sober scientific sources. That's what I did with Caro, and that led us to a discovery of the utmost importance: In ordinary biological usage, functions have explanatory depth. That doesn't mean scientists can't be wrong about particular cases, nor does it mean that scientists always use the word "function" in the same way.

Note that, even though I care about how biologists use "function," I'm not too interested in what biologists say functions are: that is, how biologists themselves would define the term "function" if you asked them, point blank, for a definition. "Function" is rarely defined in any explicit way in biology, and I wouldn't expect biologists to be able to state the rules that govern their use of "function." Just as you can use grammar flawlessly, without being able to state the rules of grammar, one can use "function" appropriately without being able to define it precisely.

Second, in some cases I consider the way functions are used in ordinary biological practice, particularly in biomedicine and biological psychiatry. Here's an example: When biomedical researchers say that a trait is *dysfunctional*, they're often indicating, in a pragmatic kind of way, that the trait is an appropriate target for medical intervention. It's the kind of thing you might want to fix or replace. This is a fact about biological practice: It's a fact about what sorts of actions biologists think are appropriate to take upon discovering that something has a function or that it's dysfunctional.

I put ordinary biological usage and ordinary biological practice above intuitions, since I think intuitions are mainly good for limning concepts; but they still have a role to play, when they're informed by good science. To the extent that I take intuitions seriously, I take them more seriously when they're about true-to-life cases (Do clay crystals have functions? What about piles of rocks?) rather than about science-fiction cases. (If the world were created five seconds ago in its present form, would anything have functions?) Sometimes people complain that intuitions have no role in serious philosophy – but soon enough, they start expressing their intuitions on all kinds of topics (they just don't *call* them intuitions). I might as well be up front about this fact from the outset.

Where have we reached so far? Functions are more complicated than they seem. The function of a trait is not just whatever it does. To illuminate functions, we considered the fact that sometimes, when biologists give functions to traits, they purport to explain why those traits exist. We then confronted this idea with the problem of backwards causation and considered whether there are any non-theist and nonfictionalist accounts of function that could solve the problem of backwards causation. I indicated that, if functions are selected effects, then we can understand their explanatory role quite easily. I then clarified what, precisely, the explanandum of the function statement is. Function statements purport to explain either why some entity has a trait, or why the members of a collection of entities generally have a trait. I recommended that we restrict functions to features (physical, psychological, or behavioral) of individuals or groups; individuals or groups themselves don't have functions. I urged that we not confuse biological and artifact functions, but the proof of that still awaits, and I set down some ground rules for the kinds of evidence that matter for defending a theory of functions.

CHAPTER 2

Function and Selection

How do functions explain anything? How can a trait's *effect* explain its very existence? Here, I'll show that if functions are selected effects, they can play the explanatory role that biologists demand of them. In the next chapter, I'll defend the more ambitious claim that functions can only play this role if they're selected effects. If we take the explanatory depth of functions seriously, the selected effects theory, or something like it, is our only path forward. The alternative is to dismiss function's explanatory ambition as an illusion or error, as something destined to remain unfulfilled.

This chapter will move through three main phases. First, I'll sketch the basic argument for the claim that if functions are selected effects, they have explanatory depth (Section 2.1). I'll then illuminate how the theory works by contrasting selection with other processes that do not, all alone, create biological functions: mutation, drift, and design (Section 2.2). Finally, I'll fortify the theory against an onslaught of traditional criticisms, and show how they all go wrong (Section 3.3).

2.1 The Traditional Selected Effects Theory

Here's how selection enters our story. Putting the point in a skeletal way: Suppose that in the past a certain trait of some species was selected for some activity by natural selection, and suppose the trait exists now because of that fact. Then, it's true to say that some members of the species have the trait now because the trait performs the activity. The trait's performing the activity in the past accounts for the trait's persistence in the species, and hence its current existence. Selection is the force that closes the explanatory circuit. (Some might suspect semantic trickery here, but I'll put those suspicions to rest soon.)

A simple example can put flesh onto the general idea. Suppose zebra stripes were selected for deterring biting flies. What that means is that at some point in the past there was a population of ancestors of modern zebras. Some were striped, some were not. (This is a simplification. Perhaps some were only partially striped, some showed only the barest hint of stripes, and so on.) The striped ones kept the flies away, so they died less often from sleeping sickness. As a result – and thanks to the mechanisms of inheritance – stripes were maintained in the population. That is why zebras have stripes today. In a selectionist explanation, a trait's effect explains its existence, without the help of backwards causation or divine intervention.

This observation, that selection accounts for function's explanatory depth, is the root of the selected effects theory of function. This theory holds, roughly, that a trait's function is whatever it was selected for, by natural selection or some comparable selection process. Karen Neander (1983) and Ruth Millikan (1984) first formulated the theory independently of each other, though earlier philosophers and scientists had glimmerings of it (see Garson 2016, chapter 3).

The selected effects theory of function is an etiological theory, in the sense described in the first chapter. It holds that function statements are condensed causal explanations. Sometimes people say the selected effects theory is a *teleological* theory, too. Unfortunately, this term is used in multiple senses, from the mystical to the mundane. There are at least three different senses that are rarely teased apart.

First, sometimes "teleology" implies full-blooded design. That's how developmental psychologists use it when they ask about whether children are innately disposed to reason teleologically (e.g., Kelemen 1999). The term "teleology" is so wedded to design that some biologists entirely refuse to use the term, and prefer to speak of *teleonomy* instead. Teleonomy is supposed to designate the quasi-teleology of biological systems (Pittendrigh 1958, 394), a kind of teleology without divine intention or backwards causation, but nobody ever explained what this quasi-teleology was supposed to be.

Second, sometimes people say the selected effects theory of function is "teleological" in that it captures the way a trait can be *for* something: stripes are for deterring flies; the tsetse's proboscis for drawing blood. I confess I have no clear idea of what it is for a trait to be for something, nor do I see how the selected effects theory is supposed to illuminate it. I read this "for" talk as, at best, a loose or metaphorical way of conveying that stripes were selected for deterring flies; it doesn't capture some interesting, distinctive feature of the world, some elusive phenomenon that a theory of function ought to shed light on. (Now, there is a sense of "for" where we say that a trait can perform a function for one organism rather than another; for example, *T. brucei's* changing coat has a function for the parasite, not for the host. Here, "for" just designates the possessor of the functional trait – more in Chapter 3).

If I were forced to use the term "teleological," I would use it in a rather mundane way, and this is the third sense. In my preferred lexicon, "teleology," above everything else, is a style of explanation. A teleological explanation purports to explain something's existence by citing one of its effects (Ayala 1970, 8). In this sense of "teleological," selectionist explanations are, technically, teleological explanations. (They're also etiological explanations.) But what's the point of speaking this way? Why insist on somewhat obscure and ambiguous labels?

Let's resume the main thread. One might suspect that the selected effects theory only seems to capture the explanatory aspect of functions, but it really doesn't. Let S be an individual or population, T a trait, and A an activity. The explanans is, T does A; the explanandum, S has T:

Explanans: T does A
Explanandum: S has T

I said that if T was selected for A by natural selection (in S), then this is a correct causal explanation.

One might protest that in a selectionist explanation the explanans is not T does A. The explanans is T did A in the past. T's doing A, right now, explains nothing. The reason zebras have stripes is not because at this very moment, some zebras are using their stripes to ward off biting flies. Zebras have stripes because, in the past, they did that. If we keep our tenses straight, we can see that natural selection won't solve the explanatory riddle of functions.

I agree that, in a selectionist explanation, the fact that T did A in the past explains why S has T now. The fact that T does A, at this very moment, explains nothing. Still, the criticism misses its mark. The statement, "T does A," in the explanans, is not about what T is doing at this very moment (e.g., on November 18, 2017). Rather, "T does A" should be understood, as Neander (2012) puts it, in a "tenseless" way – in just the way we understand statements like "beavers build dams." The statement "beavers build dams" isn't about what some beavers are up to right now. Nor does it say that beavers have always built dams, ever since they came into existence; the statement cuts a wide temporal swath. As long as enough beavers built enough dams recently, it's right to say that "beavers build dams." That's the sense in which the explanans must be understood.

Put differently, if we replace the current explanans, "*T* does *A*," with a modified explanans, "*T* did *A* in the past," we wouldn't be moving from saying something false to saying something true. We'd be moving from saying something less precise to saying something more precise. But both statements are right. At worst, the selected effects theory solves the problem of backwards causation by a sort of temporal smudging, but that's not cheating.

In capturing the explanatory depth of functions, the selected effects theory captures the other puzzling features of function, too: namely, the function/accident distinction and function's normativity. We should pause to appreciate how remarkable this is. It is as if when we use the selected effects theory to "buy" the explanatory power of functions, the normativity of functions, and the function/accident distinction, are tossed in for free.

Take the distinction between functions and accidents. Why is the nose's function to help us breathe rather than hold up glasses? Simply this: Noses are there because they help us breathe, not because they hold up glasses. That is what they were selected for. The difference between a function and a lucky accident is that the former explains the existence of the trait in question, through a natural process of selection, and the latter doesn't. For the selected effects theorist, that's all the distinction amounts to.

This theory also explains function's normativity. How is it possible for a trait to dysfunction? A trait is dysfunctional when it cannot perform its function – everyone knows that, but there's a deeper puzzle. How can a trait *possess* a function it cannot perform? By virtue of what does a function linger, like a ghost, even when the corresponding ability has vanished? The selected effects theory says that having a function depends entirely on a trait's history. It depends on things that happened in the past, perhaps long ago. Whether or not a trait has the ability to perform that function now doesn't depend on the past. It depends only on present-day facts: structure, organization, environment. It's easy to see how a trait can have a function it cannot perform: how dysfunctions are possible. I'll come back to this theme in Chapter 8.

Not all theories of function can account for the possibility of dysfunction. Consider Cummins' (1975) early account of function. He said that for a trait to have a function *F*, it must be capable of doing *F* (p. 757). Functions are capacities. That implies that as soon as a trait loses the capacity to do *F*, say, through impairment or disease, *F* is no longer its function – thus, there is also no dysfunction. Wright's (1973) view contained the same flaw. He thought that for a trait to have the function

F, it must actually have *F* as a consequence (p. 161). He recognized that, strictly speaking, this condition makes dysfunction impossible; dysfunction, he said, "offends the letter of this analysis" (p. 167). He considered ways of bending the analysis to let dysfunctions in but stopped short of revising his canonical formula. Normativity isn't always free.

2.2   Mutation, Drift, Design

To convey how the selected effects theory works, it helps to think about processes that don't make new functions all by themselves. Three such processes come to mind: mutation, drift, and human design. Consider mutation. Suppose a population of flour beetles is exposed to a harsh new pesticide and, over time, a beetle is born with a gene mutation that gives it immunity. Some would say that the gene's function, in that beetle, is to give it immunity. The selected effects theorist disagrees since selection hasn't yet taken place.

Some philosophers think this is a terrible consequence for a theory of function. Surely, the very first time a trait benefits an organism, that benefit is its function? Some have such a strong aversion to this consequence that they dismiss the selected effects theory outright or seek to minimize its role in biology (e.g., Bigelow and Pargetter 1987, 195; Walsh and Ariew 1996, 498; Schlosser 1998, 304). Some go as far as to say that biology is on their side. In other words, they say, as a rule, biologists "don't take history into account" when they give functions to traits. This is a sociolinguistic claim, and I think it's wrong. Not only are there counter-examples (think Caro), but more importantly, even if a scientist doesn't explicitly take history into account when giving out functions, he or she might implicitly do so. One goal of philosophy is to unearth the implicit assumptions that guide scientists in how they think and act (see Chapter 9).

It seems to me that, even if one doesn't accept the selected effects theory, there's a good independent reason for not giving functions to novel traits. Most philosophers accept, unquestioningly, that functions differ from lucky accidents. They agree that a trait can have a useful effect that isn't its function. Suppose the members of a religious cult abduct me, and just when they are about to sacrifice me, they notice a birthmark on my shoulder that faintly resembles one of their religious symbols – and they let me go. Still, the birthmark has no function; this was a lucky accident. Why can't pesticide resistance in a single flour beetle be a lucky accident, too?

Genetic drift is another process that can't, all by itself, create new functions, even when the trait happens to benefit the population. Consider a thought experiment suggested by Dover (2000, 41). A trait arises by a gene mutation at time $t1$. It gets fixed in the population by genetic drift by time $t2$. At some later time, $t3$, the environment changes in such a way that had the trait not gone to fixation, everyone in the population would have died. (To have a concrete image in mind, suppose there's a population of flour beetles that had never been exposed to pesticide. At $t1$, an immunity-conferring mutation arises; it's neither helpful nor harmful. It spreads through the population by drift, and goes to fixation by $t2$. At $t3$, a pesticide is introduced that would have eliminated them all but for the happy mutation.) Does the trait have a function? The selected effects theorist says "no." Some find that counterintuitive, but I don't. If a single instance of a trait can have a useful effect that isn't a function, why can't a whole population have a trait like that?

Human design is another process that doesn't create new biological functions. Rather, design creates *artifact* functions. I don't think of artifact and biological functions as two species of the same genus, like lions and tigers are two species of *Panthera*. I think the term "function," when it appears in "artifact function," just means something different than "function," when it appears in "biological function" (though see Kroes and Krohs 2009; Piccinini 2015, chapter 6; Maley and Piccinini 2018, for more on biological and artifact functions).

Here's why. One observation that others have made, such as Wright (1973, 153) and McLaughlin (2001, 45), is that artifact functions seem to depend on *intended or designed effects*, rather than actual effects. (Artifact functions are more complicated than this; intention isn't always enough, and sometimes it's unnecessary. My point is that intention is the most natural starting point for thinking about artifact functions.) Consider Lester Wire, who invented the first electric traffic light. It seems to me that the very first traffic light he made, complete with working red and green lights, had the function of regulating the flow of traffic, since that's what he meant for it to do, but that effect didn't cause its existence. The traffic light existed because Wire thought, or surmised, or anticipated that it would regulate the flow of traffic. That is the crucial difference between biological and artifact functions. With biological functions, an actual effect of the trait explains the trait's existence, not a merely presumed or anticipated or desired effect.

Some have tried to bridge the divide between artifact functions and natural functions by arguing that artifacts, like biological traits, get their functions from a selection process, but it is a virtual selection process that goes on in the designer's mind (Wimsatt 1972, 14; Griffiths 1993, 419; Millikan 2004, 11). I agree that something like that happens, but that doesn't make biological and artifact functions the same. The difference, again, is that for artifacts, at least at the outset, their effects do not explain their existence. One cannot say, of the very first traffic light ever made, that the fact that it regulates traffic explains why it exists: The fact that someone anticipated or surmised that it would regulate traffic explains why it's there. The feature that makes the selected effects theory so plausible, and nearly irresistible in the biological case, is missing for artifacts.

To be sure, sometimes artifacts do go through something like selection (Griffiths 1993, 420). Consider how cell phones "evolved" over the last several decades. They used to be large and clunky and now they are slim and lightweight. We may think of this transformation as the result of a sort of competition, in which consumers "selected" the lighter, smaller phones over the heavier, clunkier phones. Some anthropologists have used the changing functional and ornamental features of Polynesian canoes as an example of this sort of cultural evolution at work (Rogers and Ehrlich 2008), but we shouldn't make too much of the analogy. After all, an artifact, like a cell phone or a traffic light, has a function on its very first appearance, long before this sort of consumer-driven "selection" happens. I agree, however, that artifacts can acquire new functions in this round-about way; they'll just be superimposed on the ones they received from their designers.

Sometimes organisms, or their parts, acquire artifact functions, superimposed on the biological ones. For example, if an organism is intentionally modified, it comes to possess an artifact function, in addition to whatever biological functions it has. The GloFish®is a genetically modified florescent pet. In the 1990s, a scientist put genes from a sea coral into a zebrafish, yielding a zebrafish with brilliant florescent colors. The florescent color, in the GloFish®, doesn't have a biological function, since natural selection didn't put it there, but it does have an artifact function – to be aesthetically pleasing. Accordingly, even though artifact and biological functions are different sorts of things, some artifacts undergo something like selection, and some biological entities have artifact functions, but that shouldn't blur the distinction between them.

2.3   Criticisms of the Traditional Selected Effects Theory

Before moving on, it's worth fortifying the traditional selected effects theory against common objections, lest someone think there are severe problems that take it out of the running as a serious candidate for understanding functions. There are four that demand our full attention. The first is that the selected effects theory, contrary to appearance, doesn't actually make sense of function's explanatory depth. A second is that with the selected effects theory dysfunctions are impossible. Since it would be devastating for the theory if either of those were right, it's worth seeing clearly why they aren't. A third takes aim at the popular "recent history" formulation of the theory, while a fourth draws upon various intuitions to show that the selected effects theory is too narrow, or too broad. It gives functions to things that don't have them and withholds functions from things that do. (If you are not interested in seeing how the selected effects theory can respond to various criticisms, and just want to get on with the positive view, I suggest skipping directly to the next chapter.)

The Selected Effects Theory is Not Really Explanatory

Cummins (1975) argued that the selected effects theory doesn't actually give explanatory depth to functions. That's because natural selection, he thought, doesn't explain why organisms have the traits they do. Here's one way of putting the case. Consider a particular zebra; we'll call him Amadi. Why does Amadi have stripes? It is assuredly not because "natural selection gave him stripes." Instead, Amadi has stripes because of the rich interplay of his genes and his formative environment. But why does he have those genes? The reason he has those genes is because he inherited them from his parents. His parents, in turn, inherited those genes from their parents, who inherited them from their parents, and so on. If we trace the passage of those genes far back enough, we'll arrive at one of Amadi's very distant ancestors, who acquired those genes through a lucky mutation. The rub here is that when we give a fine-grained account of why Amadi has stripes, natural selection never enters our story. It's neither a proximal cause, nor a distal cause, for Amadi's stripes.

If natural selection doesn't explain why Amadi has stripes, what on earth does it explain? It explains why stripes are so common among zebras today. It explains a statistical fact about zebra populations, rather than a fact about any one zebra, just like entropy explains the behavior of large groups of particles, and not any particular one. Sober (1984) made the same point when he distinguished "variational" and "developmental" explanations. While natural selection gives a variational explanation for why most zebras have stripes (rather than being, say, monocolored), it doesn't give a developmental explanation for why any particular zebra has them.

Let's suppose that Cummins and Sober are right about natural selection. Natural selection doesn't explain why Amadi has stripes; it explains why zebras, generally, have stripes. Why would that matter? Why would that mar the selected effects theory? It would matter if one assumed that functions are always meant to explain facts about specific individuals – but why think that? As I noted in Chapter 1, a statement like "the function of zebra stripes is to deter flies" is most naturally read as an attempt to explain a fact about a population – namely, why most zebras have stripes, and not, in the first place, why this or that zebra has stripes. And everyone agrees that natural selection can explain the former fact.

True, sometimes function statements are primarily about specific individuals. The function of Gemini's standing on his hind legs is to get a piece of celery. Kainaat's finger-tapping has the function of reducing anxiety. Gerry's aggressive outbursts have the function of keeping people at a safe distance. These are bona fide function statements about specific individuals, but it seems to me that in the ordinary evolutionary context, function statements are usually attempts to explain something about populations – why some or most of the members of a population have a trait.

Incidentally, Cummins's and Sober's claim that natural selection can't explain facts about individuals has created a minor and somewhat technical literature. Neander (1988; 1995a) argued, contrary to Cummins and Sober, that natural selection *can* explain why an individual has a trait, and philosophers of biology rapidly aligned themselves with one side or the other. The Sober/Cummins view that natural selection cannot explain the properties of individuals is now called "the negative view," since it depicts natural selection as something like a sieve that merely eliminates unfit varieties. Neander's claim that natural selection can explain properties of individuals is "the positive view," since it emphasizes natural selection's creative role in producing new traits.

Neander (1995a) argues that natural selection doesn't merely explain the spread of new traits in a population. It can also explain their origin. That's because natural selection can increase the chances that a useful gene combination will arise. Putting the point simplistically, suppose there's a gene sequence $<A^*,B^*>$ that codes for a favorable trait, like opposable thumbs. Suppose there's a population of organisms all of whom have the sequence $<A,B>$. Suppose a mutation creates the $<A^*,B>$ sequence. If

$<A^*,B>$ is fitter than $<A,B>$ then, all things being equal, natural selection will spread it around. In spreading it around, natural selection increases the odds that some creature will acquire the $<A^*,B^*>$ sequence. This is natural selection's creativity at work.

Some of Neander's critics agree that natural selection can account for the origin of traits in this way. They allege, however, that that doesn't support the positive view all by itself. The positive view says that natural selection can explain why a *specific* individual has a trait (e.g., why Amadi has stripes). Neander only shows that natural selection can explain why *some* individual in a population is likely to get a trait. It doesn't explain why that particular individual (say, Amadi) was the one that got it (Walsh 1998).

For better or worse, the debate has ramified considerably. Matthen (1999) and Pust (2001) showed that the clash between the negative and positive views leads into deep metaphysical waters. (If you're not interested in possible worlds and trans-world identity, you might as well skip ahead to the next section now.) The main issue turns on how to assess *counterfactuals* involving individuals. If I hadn't pursued philosophy, I would have been a world-class cosmetic surgeon on the Upper East Side. Let's suppose that's true. What makes it true? Similarly, to say that natural selection explains why Amadi has stripes seems to imply the counterfactual claim: if there hadn't been selection for stripes, Amadi would have had some other phenotype. What would make that claim true?

The most popular way to think about counterfactuals involves possible worlds (Lewis 1973). The claim that "if there hadn't been selection for stripes, Amadi would have had some other phenotype" invites us to contemplate what's happening at different possible worlds. It tells us that at those nearby possible worlds where selection acted differently (say, there was selection for being monocolored), Amadi has some other phenotype. Here is the problem: Which zebra is Amadi on those nearby possible worlds? What are the "trans-world identity conditions" for Amadi? (I'm not asking an *epistemological* question: How might one find out which zebra is Amadi on a nearby possible world? It's not as if we're going to be transported to another possible world and charged with the task of finding him. The question is a *metaphysical* one: What is it for some zebra, on some nearby possible world, to be Amadi?)

One theory that comes into play here, made famous by Saul Kripke (1980, 113), is *origin essentialism*. Origin essentialism says that I have my parents necessarily. That is, there are a lot of ways things could have turned out for me. I could have been a cosmetic surgeon rather than a philosopher (that is, there is some nearby possible world at which I am a cosmetic surgeon). Therefore being a philosopher is not essential to me; but, according to Kripke, on any nearby possible world where my actual parents don't exist, I don't exist. Having the parents I actually have is necessary to me.

If one accepts origin essentialism, then it seems that one must accept the negative view of natural selection. Consider those nearby possible worlds at which stripes are selected against. (Perhaps there's a predator that can easily spot stripes and there's no fly-deterring benefit to offset the harm.) On those worlds, the gene for stripes doesn't exist, since if it arises at all it gets eliminated rapidly. On those worlds, Amadi's actual lineage never came into being; therefore, assuming origin essentialism, Amadi doesn't exist either. It is not the case that had selection acted differently Amadi would have had some other phenotype. Amadi just wouldn't be there.

Now, one might think the best way to save the positive view is to attack origin essentialism. In other words, by attacking origin essentialism, one knocks out a supporting beam for the negative view. This is what Matthen (2002) tried to do. The problem is this: a proponent of the positive view can't just attack origin essentialism. Additionally, he or she must furnish plausible trans-world identity conditions that yield his or her preferred result. In other words, the falsity of origin essentialism doesn't imply the positive view. Helgeson (2015) makes this point clearly, though others (Lewens 2001, 595; Pust 2001, 216) made a similar claim. The proponent of the positive view has the burden of furnishing some plausible trans-world identity conditions for Amadi. So far, nobody has done so.

I prefer the negative view. However, unlike other proponents, I don't think the negative view needs origin essentialism. Rather, the proponent of the negative view should simply insist that there are no plausible trans-world identity conditions that support the positive view. My thoughts here are informed by David Lewis's (1973, 36–43; 1986) discussions of counterfactuals involving individuals. Intriguingly, Lewis himself doesn't actually believe in strict trans-world identities. He doesn't think a person exists at multiple worlds. Rather, a person has *counterparts* at multiple worlds. A counterpart is an individual who exists at some nearby possible world and who resembles me "closely enough in important respects of intrinsic quality and extrinsic relations" (Lewis 1973, 39).

Lewis rejects origin essentialism. He thinks that when it comes to deciding who my counterpart is, different similarity relations are appropriate in different explanatory contexts. In some contexts, the relevant similarity has to do with origin. Suppose there is an individual on some nearby possible world who has my exact same history (same parents and upbringing) until the age of eighteen, at which point he decides to become a cosmetic surgeon. That individual would be my counterpart on that world. In other contexts, the relevant similarity involves phenotype and life circumstance. Suppose there's an individual on some possible world who had different parents and a different upbringing but whose life gradually came to resemble mine in every other respect (career, mannerisms, appearance, and so on). On that world, he is my counterpart. There are even worlds at which both sorts of counterparts coexist. I can have more than one counterpart at a world.

Here's my point. The proponent of the positive view says that, had selection acted differently, Amadi would have had a different phenotype. That seems to imply that there is a possible world at which Amadi exists, but he has different parents and different phenotypic characteristics. That individual resembles Amadi neither with respect to "external relations" (parentage) nor with respect to "intrinsic quality" (phenotype). Fine – but then why is this creature Amadi?

Fortunately, my position on the debate between the positive and negative views is largely irrelevant for defending the selected effects theory (though it will surface again briefly in the next chapter). The problem with Cummins's argument is that it presupposes that function statements, first and foremost, are meant to explain facts about individuals. I don't accept that restriction. I think, in the ordinary context of evolutionary biology, functions are meant to explain population-level facts (why zebras generally have stripes).

The Selected Effects Theory is Not Really Normative

A second criticism says that, despite appearances, the selected effects theory can't actually explain function's normativity. It makes dysfunction impossible. Paul Sheldon Davies (2001) came up with this argument. I'll convey it through an example. Suppose Tania is born with an underdeveloped thyroid gland that cannot produce thyroid hormone. The selected effects theorist wants to say that Tania's thyroid is dysfunctional. In other words, her underdeveloped thyroid is a member of the collection of things that have the function of producing thyroid hormone, but it cannot do so. Here is the challenge for the selected effects theorist. Let F stand for the collection of things that have the function of producing thyroid hormone. If Tania's thyroid is severely underdeveloped, why does it belong in F? By virtue of *what* does it belong in F?

Davies thinks the selected effects theorist has to say something like this: Tania's thyroid belongs in *F* since it's just the kind of thing that, long ago, was selected for producing thyroid hormone. The problem, as he rightly urges, is that that claim isn't true. Undeveloped thyroids were never selected for producing thyroid hormone, because they never did that. Historically, thyroids like Tania's were selected against. By the selected effects theorists' own lights, Tania's thyroid doesn't belong to *F*; it doesn't have the function of producing thyroid hormone; it cannot dysfunction either.

I think Davies is mistaken about what selected effects theorists have to say about Tania's thyroid. He thinks they have to say Tania's thyroid belongs in *F* because it's the kind of thing that was once selected for producing thyroid hormone. Selected effects theorists don't have to say that, and, moreover, they shouldn't. The selected effects theorist only has to say that Tania's thyroid belongs in *F* because it's related, by descent, to something that was selected for producing thyroid hormone. Being related by descent to a working thyroid is all that's needed for having a function.

Put differently: For the selected effects theorist, functions are historical properties. Having a function depends on the past; it depends on one's lineage. It is like having royal blood. To have royal blood you must either be a monarch, or you must be able to count a monarch among your ancestors. Tania's thyroid has the function of producing the hormone because of its lineage, not because of anything it does.

While Davies' challenge about normativity doesn't succeed, it raises an equally tough problem. For the selected effects theorist, having a function doesn't have anything to do with a thing's powers or abilities. It only has to do with history. But then, why should scientists care about functions? For that matter, why should anyone care about them? They would be irrelevant and useless to science, which is all about discovering the causal powers of things around us. In this spirit, Mossio et al. (2009, 821) denounce selected effects functions as *epiphenomenal.* I suspect there are many who feel the same way.

We should resist this "epiphenomenalism" objection for two reasons. First, historical properties permeate science. Consider the property of being an adaptation, being a sibling, being a volcanic mountain, being an igneous rock, or being an authentic Da Vinci. These are all historical properties. Purging historical properties from science would be a wildly revisionary move (see Brandon 2013).

Why not purge them from science? What good are they? They have several benefits. Sometimes we are just curious about historical properties.

Where did that mountain come from? Are zebra stripes an adaptation? Is *Salvator Mundi* an authentic Da Vinci? Are you my brother? People are curious about what the world is like and about how it became that way. Pointing to a historical property can satisfy that curiosity, and this is part of science's mission.

True, knowing a thing's historical properties doesn't allow me to deduce its current-day causal powers, but (and this is my second point) it does let me make reasonable nondeductive inferences about those causal powers. If I know you're my sibling, I can infer that you share half of my genes. If I know that zebra stripes are adaptations for deterring flies, I can infer that they probably do so today. Even if science were only about revealing causal powers, historical properties would still contribute to its task.

Functions and Vestiges

A third problem stems from thinking about vestiges. Traits acquire functions and they lose functions, too. They become *vestigial*. Some have argued that the selected effects theory stumbles in accounting for vestiges. Consider the human appendix, widely considered to be a functionless vestige (to be more precise, a remnant of a structure in primates, the cecum, that aids digestion). Long ago, the appendix was selected for helping digestion. It would seem that the selected effects theory implies that the appendix has the function of aiding digestion *today*. After all, the idea goes, the appendix exists today because, long ago, it was selected for aiding digestion. Isn't that enough for having a function?

In response, Paul Griffiths (1992; 1993) and Peter Godfrey-Smith (1994) recommended a slight adjustment to the selected effects theory. Only *recent* episodes of selection should count toward functions. In other words, the function of a trait is what it was recently selected for; if a trait wasn't selected for recently, it has no function. The spirit of the their recent history view is commendable, but one might worry that it appears ad hoc – that is, as a device for dodging counterexamples rather than a true *theorem* of the selected effects theory – one that proceeds from its very core. Fortunately, the restriction that Griffiths and Godfrey-Smith argue for – that a trait's function is what it was recently selected for – actually does emerge directly from the theory's core. In other words, the recent history version doesn't force us to accept any new principles, over and above those that the selected effects theorist is already committed to; it simply draws out a point that was dormant in the selected effects theory.

Consider the basic rationale for the selected effects theory: the explanatory depth of function. Sometimes, function statements are causal explanations. Those explanations have the following form:

Explanans: *T* does *A*
Explanandum: *S* has *T*

We reasoned that, *if* functions are selected effects, then function statements can be correct causal explanations.

Now consider the proposition: the function of the appendix is to aid digestion. According to the selected effects theory, this is a causal explanation with the following form:

Explanans: The appendix helps digestion
Explanandum: Humans have appendices

Is this explanation correct? No, it is not. That is because the explanans is false. It is not true that the appendix aids digestion. That is because it hasn't aided digestion in our species for thousands of years. The selected effects theory, joined with some facts about digestion, implies that the appendix does not have the function of aiding digestion. All is as it should be.

Remember, the tenseless statement, "*T* does *A*," doesn't imply that some *T* is doing *A* currently, right at this moment. It cuts a wider temporal swath – but a wide temporal swath is not infinite. If *T*s haven't done *A* anytime in recent memory, then it is simply false to say *T*s do *A*. That is just how statements about doings work when we make generalizations about species (e.g., beavers build dams, kangaroo rats leap into the air, and so on).

Functions require recent selection. Can we be more precise about how recently a trait must have been selected for to have its function? I don't think we can give any principled cutoff here, though some have tried. Griffiths (1993, 417) says that for a certain trait and a certain function, "recent history" is measured by how much time we would expect to pass before the trait atrophies, assuming it makes no contribution to fitness. This, in turn, depends on the underlying mutation rate. A problem with this clever solution is that it applies only to natural selection acting on genes. It doesn't apply to other selection processes, like antibody selection, or neural selection, which the selected effects theory should include. All we can say is that *T* must have been selected for *F* recently enough for the statement "T does F" to be true. This stretch of time will expand or contract, accordion-like, depending on the kind of selection process at issue.

Schwartz (1999; 2002) thinks the recent history view has a fatal flaw. The problem is based on a certain intuition about how functions persist over time. Suppose that, long ago – farther back than recent history goes – a trait was selected for some activity, and became fixed in the population as a result. Suppose the trait has been contributing to the average fitness of the population ever since. Suppose furthermore that by sheer statistical fluke there has been no variation in the population for that trait, from the time it was fixed in the population until now. No mutations happened to arise between the time of fixation and the present moment. Intuitively, he says, the trait still has that activity as its function, but since there is no recent selection for the activity, the recent history theory is false. Kraemer (2014) argues that such cases are empirically plausible, that is, they have a nonnegligible chance of occurring.

I don't think there's a genuine problem here. Set aside the question of how likely such "no-variation" cases might be. As a purely conceptual matter, if a useful trait was selected for long ago in the deep past but hasn't been selected for recently – it just managed to persist in a population for eons by sheer statistical fluke – I wouldn't say it has a function. I would say that the benefit it confers is a lucky accident, just like my nose's ability to hold up glasses, or a new mutation that confers pesticide resistance in a flour beetle. It would be an improbable run of good fortune.

Let me try to support my judgment with reasons. For the selected effects theorist, whether a trait's activity counts as a function or a lucky accident simply amounts to this: Does the activity explain the trait's existence, or not? If so, it's a function; if not, it's an accident. If a trait underwent selection long ago, and was fixed in a population, and was never selected for again, then it's not true to say that the trait exists now because of what it does. The trait exists now by sheer inertia. It's fine if the critic has a different way of thinking about functions and accidents, and has a good reason for saying that the activity in question is a bona fide function, and not an accident, but it would be good to see that reason. In other words, how does the critic purport to make sense of the function/accident distinction so that they achieve their desired result?

A Bevy of Counterexamples

Here's a fourth and final criticism. The idea is that there are mismatches between the things in the world that have functions and the things the selected effects theory *says* have functions. There are two prongs to the charge. The theory, it is claimed, withholds functions from things that have them and gives functions to things that don't. I'll take each in turn.

First, it withholds functions from things that really have them. Consider the very first time a zebra used its stripes to deter flies. The theory says that at that particular time and place the stripes did not have the function of deterring flies, since stripes hadn't been selected for that end. This consequence, critics allege, is counterintuitive. Surely, the very first time a zebra used its stripes to deter flies, deterring flies was its function? I don't see the problem – as I've already said, this strikes me as a good example of a lucky accident, not a function. As I see it, what makes something a lucky accident, and not a function, is just that while the trait has a benefit, it's not there because of the benefit. Again, my opponent is free to protest the way I think about functions and lucky accidents, but it would be good to recommend and defend an alternative way of thinking about the distinction.

A second problem along these lines comes from reflecting on swamp creatures. Suppose there's a lightning storm and, randomly and improbably, a bunch of molecules come together to form a molecule-for-molecule duplicate of me. It has a working heart, brain, liver, and so on. It walks and speaks just as I do. The selected effects theorist says that its heart has no function, since it was not selected for. It has the wrong history – or no history. Some find this counterintuitive, but my view is that the heart's beating, in this creature, is an accident and not a function. (There are tricky issues about whether we should call its organ a heart, but I'll set those aside.)

At any rate, I suspect that science-fiction examples can be tough for us to think clearly about. Intuitions get shaky when they're forced out of their familiar territory. First, even if I acknowledge that, *per hypothesis*, my swamp duplicate is randomly generated, it is hard to shake the feeling that it's the result of design and a miraculous intervention into nature. As such, its parts would seem to have artifact functions at the very least. Second, even though its organs do not actually have biological functions, they function just like mine. How can a swamp creature's heart function just like mine, without having a function? Because the idea of functioning like, or *functioning as*, is not a historical idea. A paperweight can function as a doorstop, but it does not have the function of being a doorstop, because it was not designed for that (Millikan 1989a, 293). A swamp creature's heart functions like mine without having any functions.

Turn to the second prong of this mismatch argument. Critics say the selected effects theory gives functions to things that don't actually have them. Bedau (1991) complains that the selected effects theory forces us to the absurd conclusion that the parts of clay crystals have functions. He argues that clay crystals undergo something very similar to natural selection. There is differential reproduction with inheritance. Still, clay crystals don't have functions. Schaffner (1993, 383) developed a similar example involving ball bearings undergoing selection by a cloner machine but that don't have functions. I'll stick with the clay crystals, but what I say of them carries over to other examples along those lines.

Some people, like Ruth Millikan (1993, 116), are willing to embrace the seemingly absurd and maintain that parts of clay crystals do have functions: "If crystals can have functions, as well as learned behaviors, artifacts, words, customs, etc., that is fine with me." I'm not willing to go with her on that one. I don't think clay crystals actually undergo selection. For a group of entities to undergo selection, those entities must belong to a single population (see Chapter 6). Moreover, for a collection of individuals to count as a population, they must have fitness-relevant impacts on one another. The things I do must affect your fitness, or vice versa, or the fitness of a third party who affects your fitness, etc. In Bedau's example, there are no fitness-relevant interactions between the clay crystals – no selection, nor any functions either. I'll return to the topic of populations in Chapter 6, where I can spend more time defending the underlying concept of populations that leads me to reject Bedau's claim.

Let me sum up. I argued that if functions are selected effects, they can play the explanatory role that biologists want them to play. They possess explanatory depth. In the next chapter, I'll defend the more ambitious claim that they can only play this role if they are selected effects. I also argued that, in order to play this explanatory role, functions must be those effects that were recently selected for. I defended the view against various objections. The most important of these objections are that it cannot make sense either of the explanatory or of the normative dimensions of function. If either of those objections were on target, it would be pretty devastating for the selected effects theory. Fortunately, they are not.

CHAPTER 3

Feedback and Functions

The selected effects theory restores explanatory depth to functions and dodges the twin pitfalls of theism and fictionalism along the way. Is the selected effects theory unique in pulling off this feat? Or are there other plausible theories of function that do the same? In this chapter, I'll scrutinize two other theories that claim to give explanatory depth to functions without relying on selection: the organizational approach and the weak etiological theory. I'll show how each comes up short. If one takes function's explanatory depth seriously, the selected effects theory has no equal.

Here's an overview of what will happen. First, I'll summarize the case for why purely forward-looking theories can't make sense of function's explanatory depth (Section 3.1). Then I'll turn to two theories that recognize that functions have an historical aspect but point to the wrong sorts of historical facts, and therefore can't make sense of explanatory depth either (Section 3.2). Section 3.3 will survey three different versions of the organizational theory of function. While these could, in principle, make sense of the explanatory aspect of functions, they fail for other reasons – either because they're too liberal or they defy the demands of naturalism. Finally, I'll consider the weak etiological theory, and show why, contrary to initial appearance, it doesn't make sense of what functions explain (Section 3.4).

3.1 Forward-Looking Functions

Before confronting the alternatives, it's worth remembering why "forward-looking" theories are out of the question. Here's what I take the core puzzle of function to be: How can an effect of a trait play into a causal explanation of that trait's very existence? Immediately, we can see that any theory that doesn't appeal to history is useless to us. No purely forward-looking or ahistorical theories can do the job. That's because a causal explanation for a current event must refer to things that happened in the past. (We encountered an exception to this general rule, but we put it to one side since it didn't apply to the usual biological examples.)

Advocates of forward-looking theories might protest against the way I've set up the problem of functions. Maybe I've misstated what functions are supposed to explain – their explanandum. Walsh and Ariew (1996, 499), for example, tell us that sometimes, when we give a function to a trait (e.g., the zebra's stripes), we're just trying to explain why the trait is likely to persist far into the future. We're not trying to explain anything about the present. This interpretation seems to run against the way mainstream biologists like Caro, Purves, or Eddy talk about functions.

To be sure, Walsh and Ariew acknowledge that sometimes functions are causal explanations for present-day traits, but they think their relational theory of function accounts for that. Roughly, they think traits don't have functions considered all by themselves but only in relation to a "selective regime"; a trait's function is just how it contributes to fitness in that regime. Here is their crucial claim: If the selective regime occurred in the past, then when we give a function to a trait, and we make the function relative to that past regime, we're explaining why the trait exists now (pp. 501–502). Strictly speaking, however, their view only implies that we can explain why a trait exists now by referring to a past function. For example, the fact that stripes had the function of deterring stripes long ago explains why they have stripes now, but that's not how biologists speak of functions. Sometimes, when biologists give a present-day function to a trait, they purport to explain the trait's present-day existence. No forward-looking theorist can make heads or tails of that explanatory ambition. Those theorists have to reject it as an error or illusion.

Some forward-looking theorists have adopted a more radical strategy for coming to terms with explanatory depth. They agree that functions are explanations of a sort; they're just not *causal* ones. Explanations fall into different categories: mathematical, reductionistic, statistical, and so on. Maybe function statements do explain the existence of traits but after the manner of a noncausal explanation instead.

I can understand how one might give a noncausal explanation for Fermat's Last Theorem, or for the sun's elliptical orbit, but how does one give a noncausal explanation of the zebra's stripes? We can get a helpful clue by glancing back upon Ernest Nagel's view about functions. For Nagel, one very important kind of explanation – in a sense, the pattern for all explanation – is what he called the "deductive model" (and what Hempel called the "deductive-nomological" model). In this model, one explains an event by deducing the event's occurrence from some laws of nature and background conditions (Nagel 1961, 403). Crucially, explanation isn't tied to causation; it's tied to deduction. To explain why zebras have stripes is to deduce that fact from some laws of nature and background facts. Those laws and background facts need not essentially have to do with the past. For Nagel, function statements are just condensed deductive inferences.

Despite its ingenuity, the deductive-nomological theory hasn't withstood the test of time. Consider a famous counterexample, proposed by Sylvain Bromberger (1966, 92): If one knows a flagpole's height, and the angle between the sun and horizon, one can deduce the length of its shadow. This exercise in deductive reasoning also doubles as a good explanation of the shadow's length. The reason the shadow is, say, 16.8 feet long, is because the flagpole is about 20 feet tall and sun's angle is 50 degrees. It can also be turned around: If you know how long the shadow is, and the angle of the sun, you can deduce the height of the flagpole. Trigonometry doesn't care what is deduced from what. The length of the shadow, however, doesn't explain the flagpole's height. If someone asked, "Why is the flagpole 20 feet tall?" and I said, "Because the shadow is 16.8 feet, the angle between the sun and the horizon is 50 degrees, and the laws of trigonometry are as they are," that would be the wrong kind of answer. Some philosophers use this example to show that explanation has something crucial to do with causation: The reason the flagpole explains the shadow but not vice versa is because the flagpole causes the shadow and not the other way around (Salmon 1989, 104).

Despite the failure of Nagel's theory of function, opponents of the selected effects theory have worked hard to revive Nagel's idea that function statements are noncausal explanations for traits. Valerie Hardcastle says there are two ways of answering a question like "Why do zebras have stripes?" The first is to identify what *causes* stripes; this story has to do with genes, environments, and ancestry. The second is to identify how stripes *benefit* zebras: "we could explain *o*'s having *T* in virtue of some advantage that *T* will give or currently gives to *o*" (Hardcastle 1999, 30). She thinks that when we state a trait's function, we cite a benefit of its being there, and merely citing the benefit counts, in and of itself, as a special kind of "explanation" of the trait.

Carl Craver makes the same sort of claim. He thinks that one acceptable way of answering a "why is it there" question is to point to how the trait contributes to some capacity of the bigger system. He calls these "contextual explanations" and says they are "legitimate answers to a second reading of the question 'Why is E there?'" (Craver 2013, 155). Arno Wouters, too, thinks we should use the term "explanation" liberally; he says there's a special kind of noncausal explanation in biology called a "viability explanation" that is closely tied to functions. Viability explanations show why an organism has a trait by pointing to a benefit of the trait (Wouters 1995, 437) but not by citing a cause of the trait.

Speaking for myself, I have a hard time seeing how pointing to a benefit of a trait can count, in any meaningful sense, as an explanation for the trait's existence, unless there's also some tacit appeal to selection underneath it. I don't want to insist, however, on any particular analysis the term "explanation." More to the point, those theorists, like Hempel, must deny that function statements are *causal* explanations for traits. In this, they clash with ordinary biological thought. As noted in Chapter 1, when Tim Caro and his colleagues said that the function of zebra stripes is to deter flies, they wanted to give a causal explanation for why zebras have stripes. If a theory manages to preserve this explanatory ambition, without inflicting violence upon it, it's a better theory – all things being equal.

3.2   Invoking the Wrong History

Set aside forward-looking theories for now. We can also rule out a certain class of backward-looking theories, too. There are some theories that invoke history, but not causal history. Since these theories don't appeal to causes, they can't make sense of how functions statements are causal explanations, either. The sorts of historical facts they look at are in the right direction, but they're still of the wrong kind for understanding functions.

Here are two examples. I'll start with Christopher Boorse's biostatistical theory of function (1975; 1976; 1977; 2002; 2014). In his view, roughly, a trait's function is its species-typical contribution to survival and reproduction (Boorse 1975, 57). The function of the eye is to see because that's how eyes typically help people live. On the surface, this is a purely forward-looking view; history doesn't raise its head.

If we dig a bit deeper into the biostatistical view, however, we find traces of history. For when we assess a trait's species-typical contribution to fitness, we must assess its behavior over a period of time that stretches back into the past (Boorse 2002, 99). Boorse's biostatistical theory is decidedly not ahistorical, if that means it expunges all reference to history. (It also means that, contrary to rumor, Boorse's theory can't give functions to freshly generated swamp creatures or instant lions.) Note that history, however, needn't be causal history. Eyes have the function of seeing because that is how they've helped us for a long time, regardless of whether seeing explains why they exist now. For Boorse, even if eyes emerged by genetic drift, and persisted through sheer statistical fluke, they would still have the function of seeing. So, while Boorse's theory includes an historical element, it doesn't turn functions into causal explanations.

Like Boorse, Bigelow and Pargetter (1987) developed a theory of function that, while generally classified as forward-looking, carries us into the past. In their view, a trait has a function if it is disposed to confer a higher fitness on the organism in that organism's natural habitat (192). What then, is an organism's natural habitat? The very idea of a natural habitat is historical, as they seem to admit; it's the environment within which the trait most recently flourished. For example, they say that if the environment changes suddenly (say, a polar bear is moved from the Arctic to the San Diego Zoo), the zoo doesn't suddenly become its natural habitat (ibid.). Natural habitats have historical depth. Again, we discover history buried within a theory that, on first glance, made do without it. Still, since the kind of history they lean on isn't causal history, it doesn't bring us closer to showing how function statements can be causal explanations.

3.3 Organizational Functions

There is a class of theories, however, that does refer to causal history, but doesn't invoke selection. Such theories take aim at the very heart of the selected effects theory, because they take function's explanatory depth seriously but cast aside selection as inessential to understanding it. Selection processes are, at best, merely one route nature uses to make new functions but not the only one. There are two such theories on the market: the *organizational approach* and the *weak etiological* theory.

The organizational approach to function is not a single theory but a collection of them. What ties them together is the idea that functions are contributions to self-persistence. Roughly, a trait's function is whatever it does that helps it persist, even when there's no selection. Consider my heart. My heart beats rapidly; that pushes blood around my body; that brings valuable nutrients to my cells and eliminates waste; I need all this to happen to stay alive. Notice that in replenishing cells with nutrients, my heart replenishes its own cells. My heart benefits from its own behavior. There is a simple feedback loop that connects the heart's pumping to its own continued survival. For the organizational view, this cycle of self-renewal is enough for new functions.

The theory can solve the problem of backward causation without selection. For consider: Why do I have a heart, right now? Why is it there? It is there because, in the past, my heart pumped blood, and by pumping blood it rebuilt itself, feeding its own cells and tissue, again and again – my heart's doings, that is, its recent doings, explain its existence. There is a cyclical causal loop at work that has nothing to do with selection. Selection needs variation and difference: One thing reproduces more than something else. Feedback loops don't require differences.

Organizational theorists know that not all functions work the same way. Not every functional trait token causes its own continued survival. Some trait tokens, like bee stingers, destroy the organism but save the lineage. Organizational theorists can adjust their theories to accommodate those facts. Schlosser (1998), for example, embraces a disjunctive theory. In his view, a function of a trait token is an activity that either contributes to the persistence of that token or to the multiplication of that kind of token.

Organizational theories come in different versions, depending on how they relate to time. They can be backward-looking, forward-looking, or atemporal. A backward-looking version says a trait's function depends on what it did in the past, that helped its persistence or reproduction (e.g., McLaughlin 2001). A forward-looking version says that a trait's function depends on what it does now that contributes to its future persistence or reproduction (see Schlosser 1998). An atemporal version says that a trait's function has to do with the way the system as a whole instantiates an abstract pattern of relationships (Mossio et al. 2009). The comments that I make below apply to all versions of the theory.

The organizational approach isn't new. Larry Wright's (1973, 161) theory of function is actually an early incarnation of it. Wright thinks that a trait T has function F just in case T is there because it results in F. Technically speaking, however, my heart is there because it, that is, that unique entity sitting in my chest, pumps blood. Read strictly, Wright's view implies that my heart has a function simply because it aided its own persistence. Wright thought that natural selection is merely one way that a trait's activity explains its existence but not the only one. Functions emerge from a more general process he called a "consequence-etiology" (Wright 1976, 91). For that reason, Wright's view is more of a precursor to the organizational view than the selected effects view.

The organizational approach, in every single one of its manifestations, faces a severe liberality problem. It can account for the explanatory depth of functions, to be sure, but at an unacceptable cost. It is profligate; it generates too many functions. Boorse (1976) saw this problem in his justly famous critique of Wright (1973). Though organizational theorists have wrestled with Boorse's objection, they never solved it. Imagine a hose in a lab that springs a leak, spraying chlorine gas. Whenever a lab technician rushes over to patch the leak, the technician is knocked unconscious, and the leak goes on. The leak has a certain effect, knocking out scientists, that causes its own troublesome persistence. That fits Wright's formula neatly, but the leak has no function. Wright's formula, therefore, doesn't work.

It's easy to create new counterexamples along those lines. Boorse said that obesity contributes to a sedentary lifestyle, which, in turn, reinforces obesity. There is a causal loop, but obesity has no function. Marc Bedau (1992, 786) asked us to imagine a stick floating down a stream, getting stuck on a rock, and being held in place by the backwash it creates. The stick creates a backwash that explains why the stick "is there," but it has no function. An autocatalytic reaction is any chemical reaction one of the products of which is a catalyst for the same type of reaction. Any simple autocatalytic reaction would therefore have the *function* of producing its own catalyst. That's going too far.

Considerations such as these led some theorists, like Neander (1983) and Millikan (1984), to identify functions with selected effects. The reason obesity doesn't have the function of contributing to a sedentary lifestyle is because it was not selected for doing that (Neander 1983, 103). At the very least, one would need a population of people that differ in adipose tissue; those with excess quantities of it would have to have higher fitness on account of the lifestyle it opens up.

Organizational theorists think they can get around Boorse's problem without mentioning selection. The idea is to supplement the core idea, functions as contributions to self-persistence, with some restrictive new principles. I call these "persistence-plus" theories. They've toyed with a variety of principles. Gerhard Schlosser adds a "complexity" principle. Matteo Mossio and his colleagues add an "organizational differentiation" principle. Peter McLaughlin adds an evaluative principle; he thinks functions have to do with values and "having a good." I don't think any of them work; it's like using duct tape to patch Boorse's chlorine leak. The whole apparatus needs to be replaced.

Gerhard Schlosser tries to solve the liberality problem by adding a *complexity* principle. Think about a simple, functionless, causal loop. For example, take of a stretch of so-called junk DNA, which just sits around, serving as a template for its own reproduction. It gets reproduced simply by not disrupting any vital activities. But junk DNA doesn't have the function of serving as a template for its own reproduction. Why not? Schlosser (1998, 320) says it doesn't have a function since it doesn't have enough complexity. A diagram can illustrate his thought (see Figure 3.1). The reason junk DNA doesn't have enough complexity is it can only do one thing to reproduce itself: namely, serve as a template for new copies.

In contrast, think about the system of behavioral camouflage deployed by the mimic octopus (*Thaumoctopus mimicus*). It sits at the other end of the complexity scale. The octopus can do many different things to ensure its survival. It can mimic flatfish, lionfish, and sea snakes, depending on what's happening around it. Assume, for the sake of argument, that there's a single neural mechanism, M, in the octopus's brain that causes these diverse behaviors (see Figure 3.2). This mechanism has a lot of complexity since it uses many different tools to achieve its own persistence.

At first glance, Schlosser's complexity principle is a good way to weed out Boorse-type counterexamples. After all, Boorse's leaky hose only has one means at its disposal to preserve itself – to release a noxious chemical; the other counterexamples (obesity, the stick) follow the same pattern. Of

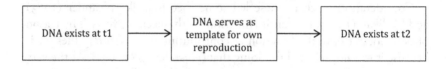

Figure 3.1   An example of a noncomplex self-reproducing system.

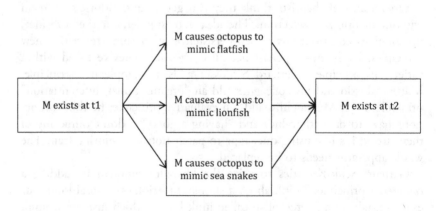

Figure 3.2   An example of a complex self-reproducing system.

course, one can always question whether the complexity principle is itself *principled*: Is there anything to recommend it other than its ability to sift genuine and bogus functions? Setting aside the issue of principle, does the solution even work?

Unfortunately, Schlosser's principle just breeds a new class of counterexamples, more complex iterations of the leaky hose. A good counterexample comes from the study of panic disorder. Panic disorder is a condition marked by recurring panic attacks, which are short bouts of intense anxiety or discomfort. Some psychologists think panic attacks have a self-perpetuating character. They set the stage for their own recurrence (Clark 1986). Moreover, they pass the complexity test, as I'll shortly explain. So, Schlosser's theory has to give functions to them. (Technically, it has to give functions to the parts of panic disorder, by virtue of how they aid the disorder itself.)

Cognitive psychologists think panic attacks perpetuate themselves in many ways (Clark 1997, 125). One way is by creating a hyperawareness of bodily sensations. After a panic attack, some people have a sharp boost in vigilance toward unusual bodily sensations. If they experience anything strange, they are prone to "catastrophize," that is, to interpret it as a sign of impending bodily harm. These misinterpretations can cause fresh attacks. There is a second way panic attacks beget new attacks. People who have a panic attack tend to avoid the places that brought it on (e.g., Odessa stops attending class because she had a panic attack there). By avoiding harmless situations, they deprive themselves of opportunities to debunk their mistaken beliefs (Salkovskis 1991). A third way panic attacks might perpetuate themselves is by inducing what psychologists call "intolerance of uncertainty" (Carleton et al. 2013). This is a kind of generalized aversion to not knowing what's about to happen.

In short, the mechanisms that promote panic attacks have a complex, self-perpetuating character (see Figure 3.3). Schlosser's view would force us to say that the parts of panic disorder have functions. For example, we would have to say that panic attacks have the function of inducing hyperalertness to bodily sensations, they have the function of inducing avoidance behavior, and so on. That contradicts the commonsense wisdom of biomedical researchers and psychiatrists, who uniformly call it a dysfunction (e.g., Ludewig et al. 2005).

Let me consider two objections the organizational theorist might raise to my panic example. First, one could insist that it's the psychiatrists who are in the wrong here, not the philosophers. In other words, maybe the parts of panic disorder do have functions; biomedical researchers and

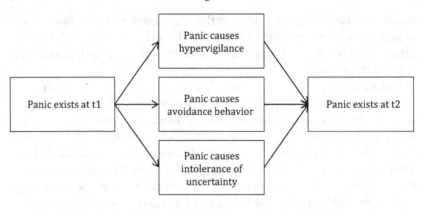

Figure 3.3   Panic disorder as a complex self-reproducing system.

psychiatrists simply mistake its true character. I sympathize with this line of thought, since I agree that psychiatrists don't have the final say in what counts as a dysfunction. Just because the psychiatrists *call* panic disorder a dysfunction doesn't mean it really is.

There are certain considerations that even seem to prop up the claim that parts of panic disorder have functions. Think about parasites, like *T. brucei*. *T. brucei* certainly has functions – the function of its ever-morphing cloak is to evade immune recognition – but those aren't functions for the host; they're functions for the parasite. Maybe the same can be said of panic disorder. These functions are functions for the disorder, not the person. If you like, you can think of panic disorder on the model of a parasite. The same point can be made of cancer cells; by replicating so quickly, cancer cells contribute to the persistence of the tumor but not the host. Unregulated cell division is functional *for* the tumor, but not for the host.

There's a big difference, however, between panic disorder and cancer cells (or parasites like *T. brucei*, for that matter). Cancer cells exhibit an extraordinary deftness in evading the normal barriers to uncontrolled replication. Those barriers include apoptosis (programmed cell death) and immune recognition. They can even change their gene expression patterns to resist immune therapy (Landsberg et al. 2012). The parasite *T. brucei*, for that matter, also shows a stunning adaptability to new situations. That extraordinary adaptability is what makes it intuitively plausible to give functions to parasites or cancer cells. (Incidentally, the reason cancer cells are so good at avoiding the normal barriers to rapid cell division is because they undergo a special kind of natural selection; because of this, the selected effects theory is comfortable giving functions to them.) That is exactly what panic disorder, however, lacks. While I'm willing to acknowledge that parts of panic disorder might have functions – I can conceive of evidence that would make me change my mind – the mere fact that it contributes to its own persistence, even in a complex way, doesn't strike me as enough.

Here's a second line of attack on the panic disorder example. Schlosser has argued (personal communication) that even if panic disorder has some functionality, there is a big difference in degree between the functionality of panic disorder, on the one hand, and the functionality of mimic octopi, on the other. Panic attacks only have a very limited number of ways to cause their own recurrence. The mimic octopus's camouflage mechanism has a much larger number of ways to secure its own persistence. If functionality comes in degrees, maybe we should say that the camouflage mechanism has a high degree of functionality, and panic attacks a low degree. That would at least allay the intuitive sting of saying that the parts of panic disorder have functions.

Since I realize that functionality comes in degrees (see Chapter 6), I am sympathetic to this line of thought, too. But I have reservations about applying it here. First, trying to count the exact number of pathways by which a trait contributes to its own persistence is a fool's errand. The problem is one of grain: How specific are we willing to get? I said there are at least three ways by which panic attacks promote their own recurrence because I used a fairly crude system of reckoning. We could always choose a more refined method of counting, where each specific mode of avoidance is a good and distinct pathway. Suppose that, on Monday, Odessa avoids jogging from fear it will trigger an attack. On Tuesday, she avoids taking a long hike in the woods with her friend. On Wednesday, she refuses to climb the four flights of stairs to her mother's apartment. How many avoidance mechanisms are we up to? One, three, or something else?

Set aside the thorny question of how to count avoidance mechanisms. Here is a bigger problem. The mimic octopus's camouflage mechanism has a number of ways of securing its own persistence, but does it have more ways of securing its own existence than panic attacks have of securing theirs? In both cases, my crude system of reckoning counted three routes. Still, the camouflage device has a clear function and panic attacks do not (or if one prefers, the camouflage device has a lot of functionality and the panic attacks little). We have to look for the difference between them somewhere else.

Here's a second version of the organizational approach, proposed by Peter McLaughlin (2001). Developing a thought experiment suggested by Adams (1999, 511), he asks us to imagine a man named Fred who develops a brain tumor. The tumor, far from being malignant, actually helps Fred because it exerts pressure on his pituitary gland, and that helps him regulate his hormones better. With time, Fred's body even treats the tumor like any other organ; it repairs and replaces its cells as it would the liver or heart. McLaughlin says that at this point, the tumor's hormone-regulation benefit has turned into a bona fide function, just as good as the heart's pumping or the liver's detoxifying, since that is how it causes its own persistence in Fred's body. Functions require *feedback* (p. 167).

McLaughlin doesn't deal specifically with the Boorse-type counter-examples, but he considers the clay crystals I mentioned in the last chapter. If you recall, clay crystals go through something very much like natural selection, but they don't really have functions. McLaughlin thinks that clay crystals reveal a deep truth about functions. The reason clay crystals don't have functions is because they don't have a *good*. Having a function means having a good. But what is it to have a good? It means that the thing can be benefited or harmed by what happens to it. Consider a zebra. Some things are good for it, like finding food and water. Some things are bad for it, like being stalked by a hyena. Clay crystals are unlike zebras because they have no "good." Nothing helps them and nothing hurts them.

McLaughlin folds all these thoughts into his theory of function. In addition to his *feedback* principle, he adds a *welfare* principle. For a trait to have a function, the trait must contribute to a larger system's good. Interestingly, Larry Wright (2013, 237) recently came to accept something like this. Now he says that in order for a trait to have a function, it's not enough that it have a consequence etiology; it must also have a "virtue etiology." I'm not sure what a virtue etiology is, but it sounds a lot like a circular causal chain that realizes a good, so it's very much like McLaughlin's proposal.

I don't know whether McLaughlin spent much time thinking about the Boorse-type counterexamples, but his evaluative stance on functions solves those, too. Consider Boorse's leaky hose. Arguably, the hose satisfies the feedback principle but not the welfare principle – therefore, no function. Unlike zebras, hoses do not really have a good. The evaluative criterion takes care of the panic disorder example, too. Panic disorder does not, itself, have a good, so its parts don't have functions. Everything is as it should be.

Unfortunately, McLaughlin says very little about what goods are, or about the kinds of things that get to have them. Does it have to do with consciousness, sentience, instinct, or something else? Whatever it is, it's something that zebras have and clay crystals don't; I want to know what that thing is. I also worry (and these worries are connected) about whether McLaughlin's view is truly naturalistic. One of the hallmarks of a naturalistic view of function (rather than, say, a normative one) is that it dispenses with any appeal to values, and more generally, anything else that doesn't fit in with the the natural scientific worldview. That's how philosophers typically understand the demand for naturalism (e.g., Bedau 1991, 647; Kingma 2010, 242), and that's how I understand it, too.

I hesitate to assert that McLaughlin's view isn't naturalistic. The point is that I don't know, since I don't know what "having a good" is. Perhaps McLaughlin can account for a creature's good in terms of a naturalistic property P, like sentience, consciousness, interests, or something of the sort. If so, we could drop all this talk of goodness and talk about P instead. We could then assess the resulting theory of function on its own merits.

It's worth pausing to ask why our theory of function should stay within the bounds of naturalism. Why demand that functions be understood in naturalistic terms? In principle, I'm not against nonnaturalism about functions, if that's what the evidence demands. In the first chapter, I entertained a plausible, nonnaturalistic theory of function: function as design. It seemed to me then, and still does now, that ideally, our theory of function should be neutral about theism, neutral about what might await us above and beyond the physical world. In addition, it should be neutral about values. It shouldn't force us to say that, in addition to all the facts that physics describes, there is an extra layer of facts, *value* facts – facts about having a good – and that this extra layer is where functions live. I want to know whether functions can find a home in ordinary physical reality. I take naturalism as a comfortable default in philosophy, but one that can be debunked if contrary evidence accumulates.

Here is a third and final variation on the organizational idea. In a series of papers, Matteo Mossio and his colleagues elaborated what they call the "organizational account" of function. (I use "organizational *approach*" to designate a family of views with a shared commitment to the idea that functions involve contributions to persistence. The "organizational *account*" is one particular theory within this family.) I'll use Mossio et al. (2009) as the canonical statement, though they've fine-tuned their view since then (e.g., Saborido et al. 2011; Moreno and Mossio 2015; Saborido and Moreno 2015.) They draw heavily upon the concepts and terminology of general systems theory, and some of their terminology varies somewhat in different places.

To approach functions, Mossio et al. (2009) first identify a class of physical systems they call "self-maintaining." Self-maintaining systems, as the name implies, are systems that do something to promote their own existence. An organism is an example of a self-maintaining system; so is a cell. For that matter, a candle flame is a self-maintaining system, as are hurricanes. A candle flame increases the ambient temperature; by doing that, it sucks in oxygen and vaporizes wax; that keeps the flame going. Now, self-maintaining systems can be broken down into parts that depend on each other, like the organs that make up our bodies, or even the molecules of a flame (this is called "closure," following Varela 1979). When a part does something that aids the persistence of the whole system, it indirectly supports its own existence since it manufactures the conditions it feeds on.

Drawing closer to functions, it's tempting to say that a function of a part of a self-maintaining system is whatever it does that helps the persistence of the whole, since that's how it aids its own persistence. The heart's pumping contributes to the life of the organism, and thereby to its own continued existence – so that is its function. They recognize, however, that contributing to a self-maintaining system is not enough for functions (Mossio et al. 2009, 825). After all, the parts of a candle flame contribute to self-maintenance, but they don't have functions, nor do the components of hurricanes. We stand in need of more restrictive principles.

The authors claim that functions belong only to a special subclass of self-maintaining systems, those that have *organizational differentiation* (826). In an organizationally differentiated system, we "distinguish between different contributions to self-maintenance made by the constitutive organization," and each component makes a "specific contribution to the conditions of existence of the whole organization" (ibid.). Simply put, an organizationally differentiated system has different sorts of parts that do different sorts of things to contribute to self-maintenance. An organism or a living cell is a good example of an organizationally differentiated system, but a candle flame isn't. All of the molecules in the flame do pretty much the exact same thing – they help raise the ambient temperature enough to suck in oxygen and melt wax.

Still, their view faces the problem of panic attacks. Panic disorders have organizational differentiation, if anything does. The system is, in the first place, a *psychological* one, made up of thoughts, feelings, and behaviors; it is physical, too, realized in the wiring of the brain. The panic system has at least three parts: hypervigilance to bodily states, false beliefs, and the attacks themselves. Hypervigilance causes the catastrophizing, which causes more attacks – which reinforces the hypervigilance. False beliefs cause avoidance behavior which causes more attacks – which reinforce the false beliefs. The components do different kinds of things to keep the whole system up and running. Still, panic disorder is a paradigm of a dysfunctional system, not a functional one. As I noted above in the discussion of parasites, while I can imagine evidence that would compel me to think otherwise, the mere fact that panic manages to persist over time doesn't seem like enough evidence.

Critics of the selected effects theory might try to turn the tables on me. They might say that the selected effects theory, too, is overly liberal and it gives functions in all sorts of counterintuitive ways (as I indicated in Chapter 2). *Tu quoque.* But once we have a clearer understanding of what selection is, all those problems disappear. I'll return to that objection in Chapter 6.

The verdict is clear: Contributions to self-persistence alone aren't enough for functions. And persistence-plus theories don't work. The complexity principle and the organizational differentiation principle don't exclude what they should, and the welfare principle strays outside of the comfortable confines of naturalism. The selected effects theory avoids all the problems of the organizational theory, so it's the better account.

3.4 Weak Etiological Functions

One final account of function deserves to be discussed. As its name implies, the *weak etiological theory* holds that, for something to have a function, it needs the right history – the right *causal* history. On this view, however, natural selection isn't required for functions. On its face, the weak theory is a serious threat to the selected effects theorist, since if it's even plausible, it would rob the selected effects theory of its uniqueness in accounting for explanatory depth. Buller (1998; 2002) is the main proponent of the weak etiological theory, although elements echo in Schwartz's (1999) version of the recent history theory of function and in Price's (2001, 40) account of function.

The weak etiological theory rests on a crucial distinction between a trait's making a contribution to fitness and a trait's undergoing selection. Fitness can be relative or absolute. Biologists usually think of fitness as relative: One organism (or trait) is fitter than another. Fitness can also be cast as absolute, where an organism's fitness is just its expected number of viable offspring, regardless of how anyone else is doing (Buller 1998, 509). In this absolute sense, a trait can contribute to fitness without undergoing selection. Imagine a population in which everyone has lungs, and everyone's lungs work equally well. Since there's no variation in lung capacity, there's no selection either. Still, lungs contribute to fitness by helping us breathe. Absolute fitness doesn't require selection. (Technically, a trait can contribute to *relative* fitness without undergoing selection either; for example, if it isn't inherited.)

Now, consider an organism with a helpful trait. Consider Amadi the zebra and his fly-deterring stripes. Here's one story about why Amadi has stripes: His parents had stripes, and stripes contributed to their absolute fitness by helping them breed, and Amadi is among their brood. This looks like a causal explanation for why Amadi has stripes, one that refers to an effect of stripes themselves. After all, if Amadi's parents didn't have stripes, they might have died long ago of sleeping sickness, and Amadi, with all his stripes, wouldn't exist. Selection, however, isn't an essential part of *this* story! The explanatory burden is entirely carried by past contributions to absolute fitness (Buller 1998, 520).

Let me put Buller's point somewhat differently. The chief job for the function theorist is to make sense of how a trait's activity can explain the trait's existence. If functions are selected effects, you can give such an account, but Buller says you don't need selection for that. The reason Amadi has stripes is because stripes are good at deterring flies (that's how stripes helped Amadi's parents live long enough to breed). The stripes' activity explains its existence by invoking the same trick that the selected effects theory does, minus the selection.

Somewhat incidentally, it's worth noting an interesting move Buller makes. He thinks function statements in biology are, first and foremost, supposed to explain facts about individuals, not facts about populations. That is, he thinks a function statement like "the function of zebra stripes is to deter flies" is supposed to explain, of any particular zebra (like Amadi), why *it* has stripes. It is not, first and foremost, about why zebras generally have stripes. In my reading, a statement like "the function of zebra stripes is to deter flies" isn't first and foremost trying to explain a fact about individuals but a fact about populations. Still, nothing will hang on this difference of opinion here.

I have some minor qualms with Buller's theory, but since none of those qualms are devastating, they're not worth repeating here (see Garson 2016, chapter 6). Here's the most serious problem for Buller's theory: It doesn't actually explain what it's supposed to. Contrary to appearance, the fact that stripes helped Amadi's parents survive doesn't explain why Amadi has stripes. It just explains why Amadi exists at all, stripes or not. Let me take a moment to unpack my claim, and then I'll give an argument for it.

Let's go back a few steps. To recap, what is function's explanatory depth? In ordinary biological usage, a statement like "the function of zebra stripes is to deter biting flies" is a causal explanation with the following form:

Explanans: Stripes deter flies.
Explanandum: Zebras have stripes.

Buller reads the explanandum a bit differently than I do; he thinks our function statement means something like:

Explanans: Stripes deter flies.
Explanandum: Amadi has stripes.

I'm not going to argue with his amendment here: I don't think his theory, however, even makes sense of that explanatory goal. At best, his theory yields:

Explanans: Stripes deter flies.
Explanandum: Amadi exists.

In other words, the weak theory can only explain why an individual exists; it cannot explain why an individual has some trait or another.

It might seem strange that certain facts in the world can explain why an individual exists, without explaining why that individual has some property. It's tempting to say: "Surely, if you want to explain why I have brown eyes, all of the facts that caused me to exist, in the first place, are explanatorily relevant to my having brown eyes. For if I didn't exist, I wouldn't have brown eyes." I don't, however, agree with that.

One way to see this is to observe that explanations have to do with contrasts, with "rather-thans." Typically, when we explain some event, we want to know why that event happened rather than some other (Dretske 1973; Sober 1984). "Why did Maitreyee invite Ashley to the prom?" What exactly are we asking about? Why Maitreyee invited Ashley, rather than someone else? Why Maitreyee invited Ashley to the prom, rather than to the movies? Or why Ashley was invited by Maitreyee, rather than someone else? By switching the contrast class, you switch the class of facts that go into the explanation. Imagine a massive sheet on which is inscribed all of the true statements about the world. Explanatory relevance is like a spotlight that shines on some statements to the exclusion of others. By changing the contrast class implicit in the question, one moves the spotlight from one set of statements to another.

If explanations are contrastive, it's easy to see why asking "why Amadi exists" demands a different answer than "why Amadi has stripes." For the contrast classes differ. In the first case, the implicit contrast is: Why does Amadi exist, rather than never having been born? To answer this, one shines the explanatory spotlight on statements having to do with viability and breeding (e.g., why Amadi's parents lived long enough to breed, why they bred with each other, and so on). In the second case, the implicit contrast is: Why does Amadi have stripes, rather than some other phenotype, like being monocolored? To answer that question, you shine the explanatory spotlight on statements about the mechanisms of inheritance (namely, Amadi inherited the genes for stripes from his parents). The fact that stripes helped Amadi's parents survive only figures into the first explanation, not the second. Sober (1995, 388–389) makes this claim about the nature of fitness. He says the fitness of an organism only explains how many offspring it has, but not the phenotypes of its offspring. In a sense, I'm just working out the implications of his point for functions.

Enough by way of clarifying the claim I'm after. Now for the argument. As I noted in Chapter 2, the most obvious way to understand explanatory relevance is in terms of counterfactual dependence. That means that X explains Y entails that, roughly, on those nearby possible worlds where X doesn't happen, Y doesn't happen either. To say that the fact that stripes helped Amadi's parents survive explains why Amadi has stripes entails that, on some nearby possible world where stripes didn't help Amadi's parents survive, Amadi wouldn't have stripes, but he'd have some other phenotype instead. And that claim is patently false.

To show that, think about the nearby worlds at which stripes don't aid fitness. There are two kinds to consider. On one kind of world, stripes don't aid fitness, but it doesn't matter: Amadi's parents get lucky and survive long enough to breed. Amadi exists on those worlds, and he has stripes, thanks to the mechanisms of inheritance. On the second kind of world, stripes don't aid fitness, and Amadi's parents die young. On those worlds, Amadi was never born, so he certainly doesn't have stripes. Either way, it's false to say that if the stripes hadn't helped Amadi's parents survive, Amadi wouldn't have stripes but some other phenotype instead.

Here's a potential objection. It echoes one that came up earlier when we contrasted the negative and positive views of natural selection. My argument in the last paragraph seems to rely on *origin essentialism*. I said that if Amadi's parents died before reproducing, then Amadi wouldn't have come into existence. But why are we entitled to assume origin essentialism? Maybe there's a possible world where, first, Amadi's parents never reproduced, but, second, Amadi was born to some other zebras and, third, Amadi doesn't have stripes.

I don't want to rest my case on origin essentialism. I don't know enough about trans-world identity to decide if origin essentialism is correct. What I will say is exactly what I said in Chapter 2: There are no plausible trans-world identity criteria (or, if you prefer, counterpart criteria) that yield the result that there is a possible world at which Amadi is both stripeless and born to different parents. For what would force us to say that that possible zebra is, in fact, Amadi, rather than some other zebra? That possible zebra resembles Amadi neither in terms of "intrinsic quality" (phenotype) nor in terms of "extrinsic relations" (parentage).

Gerhard Schlosser (personal communication) raised a different sort of objection to my claim. I say the fact that stripes helped Amadi's parents reproduce merely explains why Amadi exists but not why Amadi has stripes. He says that my claim artificially severs questions about reproduction and questions about inheritance. As he points out (I think rightly), to say that some organism reproduces implies that the organism, and its offspring, share some properties due to inheritance. As Kant would have put it, the concept of inheritance is *contained in* the concept of reproduction. You can't entirely divorce questions about inheritance and reproduction.

I agree with Schlosser's point. To say that *x* reproduces (and that *y* is *x*'s offspring) implies that *y* inherited some properties from *x*, but it doesn't specify which properties, in particular, are so shared. The fact that Amadi's parents successfully reproduced implies that they're similar to Amadi in some ways, but it doesn't imply that they're similar with respect to stripes. At best, the fact that stripes deterred flies in Amadi's parents explains why Amadi exists and why he's similar to his parents in some ways.

In short, I don't think Buller's weak etiological theory does what it claims to: namely, makes sense of explanatory depth. For, as I see the problem, making sense of explanatory depth means making sense of how a trait's effect can causally explain its existence. Applied to the zebra's stripes, this means making sense of how there can be correct explanations like *the fact that stripes deter biting flies explains why zebras have stripes*. The weak etiological theory doesn't actually do that.

Set aside the weak etiological theory. The bigger conclusion is that, if functions have explanatory depth, the selected effects theory is the only game in town. Walking away from the selected effects theory means walking away from the claim that functions are explanatory in this way. That wreaks havoc on ordinary biological usage. Let's give the biologists what they want!

PART II

Theory

CHAPTER 4

An Explosion of Selection Processes

The next three chapters make the case for the generalized selection effects theory. This theory says that a function of a trait is whatever it did, in the past, that caused its differential reproduction, or its differential retention, in a population. The first part, differential *reproduction*, captures everything the traditional selected effects theory does: the zebra's stripes, the stag beetle's mandibles, the SNAP-25 protein that causes synaptic vesicles to fuse to the neuron's wall. The second part, differential *retention*, captures selection over things that don't reproduce but just persist better or worse than others: synaptic structures, behavior dispositions, parts of clonal organisms, perhaps even problem-solving strategies. The third part, *in a population*, means that functions require the right kind of population structure. There must be a collection of entities that interact, promoting or disrupting each other's survival prospects. A key insight is that the notion of *reproduction* or *copying*, which has played such a big role in the functions debate, has created pointless obstacles and illusory problems for understanding how functions work.

In this chapter, I'll tear down the widely held prejudice that natural selection, acting between organisms, is the only kind of selection process that matters for functions. I'll begin by pointing out that traditional selected effects theorists have always admitted, at least in principle, that there are different sorts of selection processes that can create functions, including group selection and gene selection (Section 4.1). I'll then zoom in on an ontogenetic selection process, antibody selection in the immune system, and show how it creates new functions (Section 4.2). Antibody selection is a bona fide selection process, and it yields good functions, quite independently of how it helps the organism survive and breed. Then I'll turn to the problematic case of trial-and-error learning, particularly the selection of behaviors by operant conditioning (Section 4.3). A handful of philosophers have argued that trial-and-error is a valid function-bestowing process. I admire their insight in extending the selected effects theory beyond evolution, and I agree with their conclusion – but I reject the path by which they arrived there. The problem, again, is with the persistent assumption that functions need something like reproduction or copying. The next chapter develops a plausible example of a function-bestowing process that doesn't involve reproduction: namely, neural selection. Chapter Six will, at long last, present and defend the generalized selected effects theory itself.

4.1   The Breadth of Selection

Let's start from the beginning, agreeing for now that if we want to capture function's explanatory depth, the selected effects theory, or something like it, is the only path forward. What is selection? What sorts of things undergo selection? The paradigm cases of selection are traits like zebra stripes that get their functions by helping some zebras reproduce better than others. Many selected effects theorists have written as if this sort of selection between organisms is the only way new functions are created, but that is a mistake for at least three reasons.

First, biologists have vindicated Darwin in his hunch that selection can act between groups, not just between organisms. Groups can be selected over others because of differences in group-level properties. Biologists sometimes disagree about what exactly group selection requires, and they argue about how common it is in nature, but hardly anyone doubts its mere existence (Garson 2015, chapter 2). In my view, if one group is selected over another because of a group-level trait, then that trait acquires a function of its own.

Reduced sexual aggression in water striders is a terrific potential instance of a group-level adaptation; this hypothesis has been tested through ingenious multigroup experiments (Eldakar et al. 2010). Within any particular group, aggressive water striders are fitter than docile ones – but docile groups are fitter than aggressive groups, when fitness is measured in terms of total reproductive output. This clash of selection forces can, in principle, drive the evolution of docility. In this way, a group-level trait – the trait of having a high frequency of docile members – can acquire a function, even if docility is selected against at the individual level.

Here's a second reason functions don't need selection between organisms. Just as selection can work "above" the level of the organism, it can work "below" it, too. Consider the case of meiotic drive. Meiosis usually ensures that each chromosome in a pair has an equal chance of appearing in the sperm or egg. In meiotic drive, one chromosome sabotages the other and forces its way into the gamete pool for the next generation. That is a straightforward case of selection, but the "competition" is between chromosomes, not organisms. In fact, the organism often suffers from meiotic drive; the accumulation of drive genes in fruit flies and mice can cause sterility.

Still, why deny functions to chunks of chromosomes because of how they contribute to meiotic drive? The genes that contribute to sabotaging normal meiosis have a function as good as any. One may say that segregation distortion has a function *for* the chromosome, but not *for* the organism. That little word "for" doesn't allude to a mysterious principle of teleology but the bare fact that it is the chromosome, not the organism, whose fitness is in question.

As an aside, it's worth pointing out that genes aren't the only way traits are passed along. There is *epigenetic* and *cultural* transmission, too. I follow Jablonka and Lamb (2005) in using "epigenetic" narrowly, to denote physical adjustments to an individual's genome such as shifting patterns of DNA methylation. Some evidence suggests that anxiety can be passed on in just this way (see Chapter 11). For cultural inheritance, we need only consider the Japanese macaques' famous potato-washing rites (or their stone-throwing, for that matter) that spread by behavior, not genes. In theory, we can even fuse group selection with cultural transmission of traits: David Sloan Wilson (2002) treats religion as a form of cultural group selection, and Smaldino and McElreath (2016) think research labs undergo a Darwinian competition to churn out publications. Our theory of function shouldn't favor genes over other means of inheritance.

There is a third reason selection between organisms can't be the only source of new functions. There are ontogenetic selection processes, processes that don't involve the transmission of traits from organism to organism but that occur in the span of a single life; all the complex adaptations they create die with the organism. These can yield new functions too. I'll focus on antibody selection, since it is simple and unproblematic, but I'll also consider the selection of behaviors through operant conditioning.

Before turning to antibodies, however, I have to engage in a bit of intellectual positioning. Specifically, I want to position my project in relation to two other projects: first, in relation to *universal selection theory* and, second, in relation to ongoing debates about what natural selection itself is. In short, I'm not a proponent of universal selection theory, and I don't have a stance on how "natural selection" should be analyzed. Still, a few paragraphs by way of comparing and contrasting will help to bring my project into sharp relief.

For over a century, a trickle of philosophers and scientists has been fascinated by the prospect of a "universal selection theory." They maintain that selection processes permeate the living world. What Donald Campbell (1960) called *blind variation and selective retention* is, in their estimation, an all-purpose machine for producing complex adaptations, whether in nature, mind, or society. They also anticipate that the recognition of this universal force will revolutionize all of science, from biology and psychology to anthropology and even computer science. Cziko (1995) and Dennett (1996) are core contributions; Ridley (2015) is a popular attempt to revive universal selection theory.

One very provisional way to think about my project is this: I am borrowing their insight that selection processes are rife in the world, and trying to figure out what that means for functions. To my knowledge, nobody happened to notice that if selection processes are rife in the world, and if functions are selected effects, then functions are rife in the world, too. The two bodies of literature – the hunt for a universal theory of selection, and the inquiry into biological functions – never made contact. To put these two bodies of literature into conversation, however, we cannot keep the selected effects theory as it is but must modify it in its very core and carve off some extraneous elements.

That said, I want to distance myself from the aspirations of the universal selectionists. I don't think selection is as ubiquitous as they do, nor am I convinced that a scientific revolution is waiting in the wings, as soon as enough people wake up to the power of blind-variation-and-selective-retention. For example, as I will detail in the next chapter, *neural* selection isn't the only mechanism by which new synapses are formed and retained; it might not even be the most important. My only claim is that, to the extent that a certain configuration of synapses results from neural selection (and some certainly do), that configuration also has new functions, quite independent of any it acquired through evolutionary natural selection. (For that matter, I am not an adaptationist either, the kind of person who sees the hand of natural selection in practically every trait observed; I would be delighted to discover that zebra stripes have no function at all but emerged through pointless genetic drift.)

My bigger complaint against the universal selectionists is that they never succeeded in stating clearly what all these diverse processes – evolutionary natural selection, trial-and-error, cultural evolution, scientific theorizing – actually share. There was Campbell's fine slogan – "blind variation and selective retention" – but very little by way of the analytical rigor that philosophers prize. Some philosophers of science worked hard to state what exactly these different processes have in common; Darden and Cain (1989) and Hull et al. (2001) made valiant attempts in that direction. I don't think they succeeded, and they even disagreed among themselves about which processes actually count as selection. Finally, some of the applications of universal selection theory, like the idea that it can illuminate scientific theorizing, capitalist expansion, or the creation of universes, strike me as implausible on their face.

There is a second intellectual project I should contrast with my own: that's the project of capturing what *natural selection* is and how to define it. Philosophers and scientists have offered dozens of definitions of this central biological motif. Some, like Godfrey-Smith (2009) think it requires reproduction; others, like Bouchard (2008) and Doolittle (2017), think it doesn't. You might form the false impression that I side with Bouchard and Doolittle here; haven't I been saying that natural selection doesn't need reproduction? But I've said nothing about what natural selection requires or how to define it. My only interest is in functions. My claim is only that if a trait undergoes differential reproduction, or differential retention, in a population, then it gets a new function, regardless of whether one wants to call that "natural selection." In the final analysis, although I call my view the generalized *selected effects* theory, everything I wish to say about functions can be said without invoking the word "selection" at all.

4.2 Functions and Antibody Selection

Antibody selection is a process that shapes the mature immune system. It also creates new functions, as I'll shortly argue, but the selection isn't between organisms, and its products aren't passed on from one organism to another. Accordingly, the selected effects theory of function shouldn't limit functions to the products of selection between organisms.

The idea behind antibody selection is simple; Alberts et al. (2012, chapter 25), give an up-to-date review. It actually involves two distinct selection processes, not one. At birth, a mechanism of genetic recombination generates billions of genetically distinct B cells in the bone marrow. Each cell is coated with antibodies (all of the antibodies on a single cell are identical, and different cells have different antibodies). These antibodies are "Y"-shaped proteins with slightly different molecular textures. When an antibody meets a foreign particle with a similar shape (the antigen), its corresponding B cell is massively replicated. As a result, we have just the antibodies we need, in large enough quantities, to fight off the antigens that are most common in our surroundings. Niels Jerne (1955) and Frank Burnet (1959) proposed this "clonal theory of antibody production" in the 1950s, and biologists developed it systematically in the 1960s. The analogy to natural selection wasn't lost on Jerne; he called it the "natural-selection theory of antibody formation."

Hull et al. (2001, 518), in mapping the similarities between antibody selection and natural selection, point to an obvious function for antibody selection as a whole. Antigens replicate and evolve much faster than human DNA. In order to keep up with them, there has to be a rapid form of natural selection to counter them. It's just as if natural selection stumbled upon itself as a terrific solution to a design problem.

After a B cell has been massively replicated, it undergoes a second round of selection called "affinity maturation." As I said, after the antibody meets its antigen, its B cell starts replicating, but this replication process is unfaithful. That's because B cells are equipped with an enzyme, activation-induced deaminase (AID), whose job is to litter the B cells' genetic code with mutations ("somatic hypermutation"). When a B cell replicates, it creates a batch of daughter cells that genetically differ from each other and from their parent.

What's the point of all this hypermutation? Why shouldn't the B cell be allowed to create an army of clones? After all, since it did such a good job binding to the antigen, you would think that more of the same would be even better. The reason is that even successful B cells can be improved. Because of somatic hypermutation, some of a B cell's daughters will be slightly better than their parent at matching their antigen – they will have a higher "affinity" for that antigen. Those daughter cells with the highest affinity for the antigen will themselves be replicated; those with lower affinity will self-destruct. The amount of antigen circulating in the lymph nodes is a limiting resource over which the B cells compete. Affinity maturation is as ruthless a competition as any in nature.

Let's return to functions. Consider a particular B cell. Suppose this B cell produces an antibody with a specific shape, one that no other B cell produces. We will call this shape Y_{1046}. Y_{1046} is a type of shape; it can be instantiated in different tokens and can vary somewhat between tokens. When Y_{1046} first came into existence, through the random shuffling of gene segments, it had no specific function. It had a *generic* function, since it was an antibody. Its generic function was just to bind to antigens for the ultimate purpose of eliminating them from the body. But it shared this function with the antibodies on every other B cell: with Y_{23}, Y_{31056}, and Y_{106523}. All of them had the generic function of binding to antigens, owing to their origin in evolutionary natural selection.

Now, imagine this B cell comes into contact with its corresponding antigen: say, a surface protein of the parasite *T. brucei*, and on that account gets mass-produced. A second round of selection takes place, affinity maturation, yielding a daughter B cell coated with extremely high-affinity variants of Y_{1046}. What is the function of Y_{1046} now? Its function, after affinity maturation, is to bind with one of *T. brucei*'s surface proteins, since that's what caused its mass-production in the lymph node. It never lost its generic function, to bind with antigens for the purpose of eliminating them, but it now has a new function superimposed on the first. This new function is specific to it; it belongs to no other antibody type.

Someone might doubt that B cells really acquire new functions through antibody selection, but such doubts would be spurious. After all, the new function statement – "the function of Y_{1046} is to bind one of *T. brucei*'s surface proteins," is both explanatory and normative, which is what we wanted out of our theory of function in the first place. First, it is explanatory. The fact that Y_{1046} binds *T. brucei*'s surface protein explains why it exists: that is, it explains why so many B cells have Y_{1046}. It is normative, too. Recall that B cells form lineages. They stand in parent–offspring relations to one another. When a parent B cell, coated with an antibody with function *F*, produces daughter cells, each daughter cell is coated in antibodies with function *F*. This is still true even if some of those antibodies are misshapen or deformed, and cannot do *F*. As a consequence, an antibody can have a function that it cannot perform: that is, something like dysfunction is possible.

There are two potential misunderstandings that we should take pains to avoid, since they would distort the foundations of the theory I'm building. First, my claim that antibody selection creates new functions is not based on an argument by analogy. My claim is not that antibody selection creates new functions because it's similar enough to selection between organisms. It is similar, but that's hardly the point. We should accept antibody selection as a function-bestowing process for the same reasons that we accept selection between organisms as a function-bestowing process. My argument is simply an application of Mill's (1882, 212) parity of reasoning. Remember, the reason we decided that natural selection creates new functions is that it allowed us to understand three features of functions that were otherwise puzzling: the explanatory and normative features of functions and the function/accident distinction. Since antibody selection also makes sense of those features, it makes new functions too.

My second point is that the sorts of functions antibody selection creates are not merely *derived proper functions* in Millikan's sense. They are direct proper functions in their own right. It's not as if antibodies acquire new functions in some parasitic or half-hearted way. To show that, I need to say a bit more about what direct and derived proper functions are. In Millikan's view, a trait's activity is its *direct proper function* if the trait was selected for that activity by natural selection or a comparable selection process that involves copying or reproduction. Therefore, zebra stripes have the direct proper function of deterring flies if that is what they were selected for. A trait has a derived proper function F when it is related, in the right sort of way, to a trait that has the direct proper function F. What exactly must this relation be?

The standard idea (as in Millikan 1989a, 288) is that T^* has the *derived* proper function F when there is some other trait, T, that has the *direct* proper function F, and T normally carries out this function by producing items like T^*. Her stock example is the chameleon (Millikan 1984, 41–42). The pigment rearrangers in the chameleon's skin have the function of making the skin match the environment. Suppose a chameleon wanders into a new environment, one never yet encountered in the history of the species (say, a green environment with brown stripes). Suppose (just for fun) that the chameleon's pigment rearrangers actually produce a matching pattern of camouflage, c. That striped pattern of camouflage, described in its specific detail, has no direct proper function, since it was never selected for. Instead, c has a derived proper function. It has the derived proper function of camouflage because it is produced by a mechanism (the pigment rearrangers) that has the direct proper function of camouflage and that normally carries out this function by producing things like c. (To be precise, c has two sorts of derived proper functions: an *invariant* derived proper function and an *adapted* derived proper function. Its invariant derived proper function is camouflage. Its adapted derived proper function is to match the striped environment.)

Now, suppose someone thought that natural selection in the evolutionary sense is the only process that creates direct proper functions. That person might argue along the following lines: "I agree with you that an antibody mass-produced by antibody selection acquires a function, but it doesn't acquire a direct proper function, since it wasn't shaped by natural selection. Rather, it acquires the derived proper function of boosting immunity, since antibody selection, as a general-purpose physiological mechanism, was shaped by natural selection to boost immunity, and it normally carries out this function by differentially cloning certain antibodies over others."

This line of thought represents a totally arbitrary construal of the selected effects theory. It privileges natural selection – the kind that acts between organisms – as the gold standard for making new functions. What would drive someone to privilege natural selection that way? The reason we identify functions with selected effects isn't because we're obsessed with Darwin; it's because doing so solves all the big problems of biological function. By parity of reasoning, any process that can be used to solve those puzzles should also give rise to new functions, too, and not just derived proper functions but direct proper functions, as good and solid as any in the biological world. Therefore, antibody selection creates direct proper functions, *in addition to* any derived proper functions it creates.

Here's an interesting consequence of the picture I'm constructing: It's certainly possible for a specific antibody, as a result of affinity maturation, to acquire two different functions, a direct and a derived, and for these two functions to converge. Consider Y_{1046} again, which has the function of binding to one of *T. brucei's* surface proteins. Due to antibody selection, it acquires the direct proper function of binding to this surface protein, but antibody selection, as an all-purpose machine for expelling antigens, was shaped by natural selection for mass-producing antibodies that match the environment, and it does so by mass-producing antibodies like Y_{1046}. Y_{1046}, plausibly, has the (adapted) derived proper function of binding to one of *T. brucei's* surface proteins, too. In general, when it comes to antibody selection and neural selection, direct and derived proper functions will converge.

It's possible, at least in theory, that they won't – that is, that antibody selection can create an antibody that doesn't have a derived proper function. Consider the first creatures whose immune systems operated according to the principles of antibody selection. In those creatures, antibody selection as such, as a general-purpose mechanism for defending the host against antigens, didn't have a function, since it hadn't been selected for. Yet, the antibodies that were shaped by the process of antibody selection would still have had direct proper functions. They would have had direct proper functions of binding to antigens, since that's what they were selected for. I'll return to this topic at the end of Chapter 5.

4.3   Is Learning a Selection Process?

As I noted above, a handful of philosophers think learning creates new functions. Wimsatt (1972) and Millikan (1984) tried to extend the selected effects theory (or something like it) to the selection of behaviors by operant conditioning. Millikan also thought it applied to imitation. Papineau (1984; 1987; 1993) suggested that *beliefs* undergo a kind of natural selection that creates new functions. Griffiths (1993) and Millikan (2004) considered some forms of problem-solving to be selection processes that make new functions. Are these legitimate moves?

I agree that we should be able to say that trial-and-error (for example) creates new functions. After all, it makes sense of the function/accident distinction, and the explanatory and normative side of functions, as I'll explain shortly. The problem is that the traditional selected effects theory doesn't actually let us say that, since that theory restricts functions to things that reproduce. A new theory is needed. I'll walk briefly through each of these views to show where they get sidetracked.

Wimsatt (1972) noticed that there was an intimate connection between function and natural selection. He didn't define "function" in terms of selection, since he didn't think that, as a matter of how ordinary people think and talk about functions, functions were selected effects (p. 15). He thought, however, that selection processes were intimately involved in functions: "the operation of selection processes. . .appears to be at the core of teleology and purposeful activity wherever they occur" (p. 13).

For Wimsatt, selection isn't just natural selection. It refers to any process that exhibits Campbell's blind variation and selective retention. There must be a population of like entities, a mechanism of random variation, and a process by which some members persist or multiply, while others are eliminated. Trial-and-error learning surely fits this pattern, he thought. I agree with Wimsatt that selection processes, broadly understood, underlie all correct function statements, even though Wimsatt did not go so far as to define "function" in terms of selection.

Millikan also thinks trial-and-error creates new functions. In her view, the function of a trait is whatever it did in the past that brought about its own reproduction, over some alternative trait. In the ordinary evolutionary case (think zebra stripes), a trait gets reproduced over others because of its specific contribution to fitness. In trial-and-error, a behavior (say, a handshake) gets reproduced over others because it yields a reward: "Behaviors that result from training or from trial-and-error learning involving correlations of a reward with the behavior have as direct proper functions to produce that reward" (Millikan 1984, 28).

Millikan knows that trial-and-error isn't the only way that we learn. She admits it might be the exception rather than the rule (1989b, 292). Her point is only that when it happens, it creates new functions. Functions, in her view, don't require selection between organisms. Even if creationism were true – that is, even if all species were created several thousand years ago in their current form – trial-and-error would still create new functions.

Incidentally, she also thinks *imitation*, rather than trial-and-error, can create new functions (Millikan 1984, 21). If I see someone make a gesture and benefit from it, and I do the same in hopes of a reward, the gesture has successfully been "reproduced" from person to person. This doesn't deviate from the selected effects theory; we're just invoking a behavioral mode of inheritance instead of a genetic one. Papineau (2006) rightly emphasizes the importance of nongenetic inheritance in thinking about functions.

Here's the problem with Millikan's way of extending the theory. She thinks functions belong to members of reproductively established families. That means that the thing with the function must be descended from others by something like reproduction or "copying." There must be a lineage of items chained together by parent–offspring relations. The problem is this: What exactly gets reproduced in trial-and-error? Is my saying "hello" to someone today a copy of my saying "hello" to someone yesterday? Copying doesn't seem like the right relation here (Grantham 2001; Kingsbury 2008). Instead of reproduction or copying, there's just a persisting mechanism (my being disposed to say "hello") that churns out behavior tokens that resemble each other.

One obvious way of saving Millikan's insight is to say that behavior *tokens* are not being selected for but behavior *dispositions* are. That seems closer to the truth, but note that behavior dispositions don't reproduce, either. They simply persist for shorter or longer periods of time, so Millikan's theory cannot give them new functions. To be precise, her theory can give them derived proper functions, half-hearted and parasitic, but not direct proper functions.

Papineau makes a similar move in his discussion of how beliefs acquire new functions. He first characterizes beliefs "functionally": that is, in terms of their inputs, outputs, and interrelations. My belief that it is warm outside is a mental state that is triggered by warm places, and it mingles with other mental states to cause me to act in appropriate ways like taking off my coat. He then suggests that these functionally characterized beliefs can undergo a form of natural selection, in which some get fixed over others because of their useful effects (Papineau 1984, 557).

Papineau is on to something very important, but he doesn't flesh it out in any real detail. Kinsgbury (2008) complains that his discussion raises more questions than it answers. First, what exactly is being selected for – beliefs, concepts, or something else? Second, and more importantly, what is reproducing? What are the lineages? Where I am writing, it has been raining for two days. Is my occurrent belief that it is raining, a copy of yesterday's occurrent belief? Kingsbury is right that, given the theory of function Papineau seems to endorse, it's hard to see how beliefs create new functions, since they don't form lineages of reproducing entities: They either survive or perish (Kingsbury 2008, 501).

The same issue affects attempts like Millikan's or Griffiths' to think about *problem-solving* as a function-bestowing process. Griffiths (1993, 419) suggests that when people make artifacts, they consider alternative designs and select the one most suited to their purposes. This "hypothetical competition" can yield new functions. Along similar lines, Millikan (2004, 11) talks about the kind of virtual trial-and-error involved in solving tough problems. We ponder various courses of action and accept or reject them in terms of their hypothetical effects. Dennett (1996) calls us "Popperian creatures" for our ability to try out solutions in virtual space before putting them to the test, although he does not connect that discussion to the topic of functions. The problem, for Millikan and Griffiths, is that there's nothing like reproduction happening. At best, we have the differential persistence of strategies rather than their differential reproduction.

Intriguingly, Peter Godfrey-Smith, early on, puts his finger on exactly this tension – that is, the tension between wanting to extend the selected effects theory to learning, but not being able to because there's nothing like reproduction happening. In one paper (Godfrey-Smith 1992), he advances the optimistic line that learning is a legitimate selection process, and it creates new functions: "the selective approach [to defining "function"] is in no way tied to the genetic kind of biological evolution. . .A selective basis for functional characterization is available whenever learned characters are maintained within the cognitive system because of their consequences" (p. 292). This puts him in league with Millikan, Griffiths, Papineau, and Wimsatt.

Soon afterwards, however, he realized that one cannot extend the selected effects theory to learning in any straightforward way. In a later paper (Godfrey-Smith 1993), he concluded that the only way to solve the Boorse-style counterexamples (recall the leaky hose) is to limit functions to things that reproduce (pp. 198–199). The reason leaky hoses don't have functions is because they don't reproduce. An unattractive consequence of this restriction, he noted, is that learning generally doesn't create new functions, since there's nothing like reproduction. I appreciate that he recognized the problem, but there's a better way of avoiding the Boorse-type counterexamples, one that doesn't sever evolution and learning.

So far, I've simply conveyed what I take to be a standard reading of the traditional selected effects theory. This theory holds that the function of a trait is whatever caused the trait to be differentially reproduced within a population, whether the population consists of organisms, antibodies, or behavior tokens. A key problem with this theory is that it's based on the unprincipled restriction that functions belong only to things that reproduce. In the next two chapters, we'll see what happens when we drop this restriction.

CHAPTER 5

Selection and Construction

For the traditional selected effects theorist, functions belong, first of all, to things that reproduce. What happens when we drop this requirement? Then we can give functions to entities that merely manage to persist longer than others in a population: *differential retention*. I'll spend the rest of the chapter developing a plausible example of functions created this way, *neural* selection, so that we have something tangible to ponder. There are a few reasons to delve into the details of neural selection. First, it's a very plausible candidate for functions without reproduction. One benefit of GSE, which I'll set out in the next chapter, is that it lets neural selection create new functions. Second, neural selection has special significance for extending GSE to think about psychology and mind, as we'll see when we come to the topics of mental illness (Chapter 11) and representation (Chapter 12). It's worth wading through some of the details.

Here's the blueprint for this chapter. First, I'm going to introduce the bigger problem of how the brain decides which synapses to connect to which (Section 5.1). Then I'll introduce two mechanisms, among others, that the brain uses to solve that problem, neural selection and neural construction (Section 5.2). Section 5.3 will spend some time working out a detailed example of synapse selection. From there, I'll take a short detour to consider the function of neural selection as such – in other words, why did neural selection, as a general purpose mechanism for building brains, evolve? The most plausible answer is that it confers extraordinary plasticity on the developing brain (Section 5.4). Finally, I'll explain how synapse selection creates new functions (Section 5.5). In short, if one synapse is selected over another because of some effect, then the function of that synapse is to produce that effect. I'll also consider other examples of differential persistence in the natural world to show the breadth of application of my theory of function.

5.1 How to Build a Brain

For now, set aside functions; a brief primer on neural selection is in order. Neural selection refers to competitive interactions between neural structures, such as synapses or neurons. Neurons and other neural structures don't reproduce; they don't form lineages bound by parent–offspring relations. (Although some neurons can be generated throughout life, that's not a kind of reproduction.) Instead, there is only *differential retention*: Some neural structures persist and grow; others wither away or die. The traditional selected effects theory cannot give them new functions. That's one reason why we need a new theory.

In principle, there are at least three kinds of neural selection, depending on what is competing with what: synapses, whole neurons, or groups of neurons. The last of these, neural group selection, is tied to an ambitious and speculative theory of brain development and cognition, promoted by the late Gerald Edelman, who won the Nobel Prize for his work on antibodies. It's no accident that this pioneer of antibody selection was also an enthusiast of "neural darwinism,"(1987), since he glimpsed the same principles at work in both. People have criticized Edelman's theory for being overly speculative (see Barlow 1988; Purves 1988; Crick 1989). The criticism is fair, since there are no well-documented cases of neural group selection, despite its theoretical possibility. Still, any shortcomings associated with his theory don't threaten the other two kinds of selection, both of which are extensively documented, as I'll soon show. Here, I'll focus on synapse selection.

What is synapse selection? First and foremost, it's a mechanism that creates new synapses and decides which ones to keep. To understand synapse selection, we have to understand the bigger problem of how the brain creates synapses. There are about a hundred billion neurons in the human brain, and about a hundred trillion connections between them. A vast amount of our physical, cognitive, and behavioral functioning depends on how those neurons are connected together. Hence the problem: How does the brain decide which neurons to connect to which?

At the most general level, the brain deploys two different methods for putting neurons together, activity-*dependent* and activity-*independent* ones. Consider a single neuron, which can form connections with up to 10,000 others. The pattern of connections is activity-dependent when it's sculpted by that neuron's intrinsic activity (specifically, the production of action or graded potentials). The pattern is activity-independent when it's not based on that neuron's activity but on genetic signals or external chemical cues.

For an example of an activity-independent mechanism, consider Roger Sperry's chemoaffinity hypothesis (e.g., Sperry 1965). In Sperry's view, each neuron has a special chemical tag, fixed by genes or in early development, which guides the neuron to a similar tag. Its pattern of connections has nothing to do with its intrinsic activity; instead, the neuron is passively dragged along by "intricate chemical codes under genetic control" (p. 170). Sperry knew perfectly well that life experiences can "fine tune" the pattern of synapses (Meyer and Sperry 1976, 113), but he downplayed the role of experience in shaping the mature brain.

On the surface, Sperry's view is hard to swallow. How can genes alone specify a hundred trillion synapses? No doubt, plenty of synapses are hard-wired, in some creatures more than others. In the zebrafish and newt, the basic blueprint for vision relies on genetic cues, instead of visual experience – but they have far fewer neurons to work with. For example, newts have about 100,000 tectal neurons, so the task of wiring them up could reasonably be under genetic control. Sperry studied these neurons extensively in the 1930s and 1940s, perhaps overgeneralizing from those cases.

It seems like some of our synapses must be produced or maintained by activity-dependent methods instead. There are two main mechanisms in this general scheme. The first is neural selection (and, more specifically for my purposes, synapse selection) and the second is neural construction. Although selection and construction have both been demonstrated, people still argue about which one, if either, is more common.

5.2   Selection and Construction

According to the *selectionists*, the formation of new synapses is largely *random* and *exuberant*. The brain produces far more synapses than it actually needs. Over time, some of these synapses prove useful, and others do not. The useless ones are eliminated, or pruned away (hence the term "synaptic pruning"). The end result of synapse selection is a highly efficient system of neural connections. This core picture was promoted by neuroscientists such as Changeux (1985); Edelman (1987); Gazzaniga (1992); Deacon (1997); and Innocenti and Price (2005).

Synapse selection might be taken to suggest a simplistic picture of human brain development: People start life with the most synapses they'll ever have, and they lose them over time. While some early evidence suggested that frontal lobe synapse density in humans peaks in infancy and declines thereafter (Huttenlocher 1979), that doesn't appear to be the general trend (Purves et al. 1996). In fact, the growth of the brain from infancy to puberty strongly suggests a net increase of new synapses! Indeed, some have used this fact as damning evidence against synapse selection (e.g., Purves et al. 1996; Quartz 1999).

Fortunately, selectionism doesn't imply that the brain undergoes a net loss of synapses during life. Rather, this cycle of *blind variation and selective retention* can occur again and again. Synapse selection is consistent with the fact that the total number of synapses gradually shoots up through puberty, just as evolutionary natural selection is compatible with the fact that populations can get bigger over time rather than shrink.

You might think that neural selection is an inefficient and wasteful method for deciding which neurons should be connected to which. Why all that pointless demolition? The alternative to neural selection is neural construction. *Constructionists* think the brain builds new synapses on a parsimonious, "as-needed" basis. Consider Hebb's rule: "neurons that fire together, wire together," inspired by the work of Canadian neuroscientist Donald Hebb (1949), although he didn't coin the phrase. Hebb's rule is standardly read as saying that if one neuron activates another, the synapse between them is strengthened. This is an example of neural construction. There's nothing like the selectionist cycle of proliferation-and-elimination taking place. Rather, the brain strengthens synapses depending on how much they're used.

Neural construction doesn't only strengthen existing synapses; it can also build them from scratch. Suppose one neuron (N_1) synapses onto another (N_2). Suppose that, as a result of N_1 activating N_2, the synapse between them is strengthened. One way this can happen is if N_1's neuron grows new axon terminals, which make additional contacts with the target cell. N_1's axon can also send out new branches, "exploratory branches," that seek out and connect with new neurons in N_2's neighborhood (Purves 1994, 59). Here, neural construction builds brand new synapses; it doesn't just strengthen old ones.

In the 1990s, some theorists took a polarizing approach to the whole issue of construction and selection. It was easy to get the idea that there was an exciting debate between the two views, and nobody can resist a good debate. Proponents of constructivism like Dale Purves and his colleagues (1996, 463) warned against "unwarranted enthusiasm for the idea that neural development proceeds by winnowing an initial excess," which "can only obscure the essentially constructionist nature of mammalian brain development, and may impede the effort to understand it." One paper even styled itself as a manifesto for the constructionist's cause (Quartz and Sejnowski 1997).

Some selectionists were also guilty of overstating their case, with the kind of heavy-handedness that invited backlash. The neuroscientists Jean-Pierre Changeux and Stanislas Dehaene (1989, 82) once wrote, with an absolutist streak, "[synaptic] activity does not create novel connections but, rather, contributes to the elimination of pre-existing ones." Earlier, Changeux (1985, 249) made the point even more dramatically in a popular book: "to learn is to stabilize preexisting synaptic combinations, and to *eliminate* the surplus."

The idea that there is a deep debate here is misguided for three reasons. First, nobody contested the mere existence of either mechanism. Everyone recognized that there are solid examples of both. The debate was about how relatively common these mechanisms were. The problem (and this is the second point) is that nobody really knows. Few neural systems have been studied very extensively with this exact question in mind, and much of the evidence for one view or the other comes from computer simulations, which can mislead us about what the brain is actually doing. Third, there are proven examples in which both mechanisms, selection and construction, join hands. For example, the formation of abnormal ocular dominance columns in the mammal's visual cortex seems to involve both the competition-driven elimination of unnecessary connections and the activity-dependent construction of new ones (e.g., Antonini and Stryker 1993, 3572). Maybe this kind of cooperation of mechanisms is the norm, not the exception.

At any rate, I'll set neural construction to one side for the rest of this chapter, since I want to focus on how selection creates new functions. In the next chapter, I'll argue that constructionist mechanisms do not create new functions; they just amplify existing ones. Picture it like a river delta. Selection creates new branches; construction dredges them deeper.

5.3   War of the Synapses

There are several well-documented examples of synapse selection. Synapse selection is implicated in the development of the neuromuscular junction, filial imprinting, olfactory neurons, and abnormal ocular dominance columns (see Lichtman et al. 1999; Wong and Lichtman 2002; Innocenti and Price 2005 for reviews). One of the first good examples of synapse selection came from the rat's neuromuscular junction (Brown et al. 1976). At birth, each muscle fiber is connected to a large number of motor neurons (see Figure 5.1): Several motor neurons converge onto each fiber. This yields a messy pattern of connections between motor neurons and

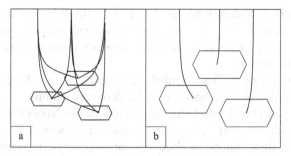

Figure 5.1   Innervation of skeletal muscle of newborn rats. The first panel (a) depicts
the multiple innervation of muscle fibers by motor neurons; the second panel (b)
depicts the one-to-one pattern of connectivity that emerges by two weeks after
birth. Redrawn from Purves and Lichtman (1980, 155).

muscle fibers (Figure 5.1a). Soon enough, motor neurons start withdraw-
ing their axons. After about three weeks of recession, an orderly, one-to-
one pattern emerges (Figure 5.1b). Without this specific pattern, fine
motor coordination would be impossible. The neuromuscular junction
supports the proliferation-and-elimination picture.

Another textbook example of synapse selection comes from the study of
ocular dominance columns in the mammal's visual cortex – specifically,
the formation of *abnormal* ocular dominance columns (e.g., Price et al.
2011; Kandel et al. 2013, 1265). I'll delve fairly deeply into the mechanics,
since we'll need to have them in front of us to make progress on functions.
(It is not clear, at present, whether synapse selection is involved in normal
ocular dominance column formation, too. More on this shortly.)

In the cat's visual system, there are at least two crucial sets of
connections that ensure that information about light is routed from
the retina to the visual cortex. First, there are retinogeniculate neurons,
which connect the retina to the midbrain's thalamus (specifically, to the
lateral geniculate nucleus). Second, there are the geniculocortical
neurons, which connect the midbrain to layer IV of the visual cortex.
The layer IV cells transmit visual information to other parts of the cortex
for more refined analysis.

Most cells – about 80 percent – in layer IV can respond to information
from either eye. They are "binocularly driven," though some show a
preference for one eye or another. The remaining cells can only respond
to information from a single eye. These are the "monocularly-driven" ones.
The extent to which a cell is driven by one eye over another is called its
"ocular dominance profile." Cells with the same ocular dominance profile

are not assorted randomly but tend to clump together (the "ocular dominance columns"). These columns can be stained with various methods
and made to appear in tissue slices as a series of alternating bands, much
like a zebra's stripes.

The neuroscientists David Hubel and Torsten Wiesel did experiments
in the 1960s that showed how the ocular dominance profile of a cell could
be modified by experience. (They shared half of the Nobel Prize in 1981;
the other half went to Sperry for his split-brain work.) In one set of
experiments, they blinded newborn kittens in one eye (monocular occlusion) by sewing the eyelid shut. Then they monitored the visual cortex for
several weeks. They found that, after occlusion, most layer IV neurons
switched their ocular dominance profile. Those that were binocularly
driven became monocularly driven, and could only respond to the nondeprived eye. Even after Hubel and Wiesel reopened the occluded eye, the
layer IV neurons weren't responsive to it anymore.

The effects of monocular occlusion can be visualized by staining. In the
monocularly occluded kitten, the ocular dominance columns have strik-
ingly different widths. Those associated with the normal eye are much
thicker, and those associated with the occluded eye are much thinner. This
arrangement works well for the kitten since it helps to maximize acuity in
the normal eye; a greater portion of cortical real estate is devoted to
processing and analyzing information received by the functioning eye.

How does the shift in ocular dominance happen? What causes a brain
cell to prefer the normal eye? Hubel and Wiesel reasoned that the shift
involves a "competition" between those neurons associated with the
normal and the occluded eye (Wiesel and Hubel 1963, 1015). In other
words, they thought that the frequent activity of some of the synapses
causes the elimination of the ones that are infrequently used. The latter do
not just atrophy from disuse; they are actively outcompeted. This becomes
clear when we look at kittens that are blinded in both eyes at birth.
Geniculocortical neurons in animals raised in complete darkness don't just
wither away from disuse, and the cortical neurons they synapse onto
remain binocularly driven (LeVay et al. 1980). Something about the firing
of one group of neurons makes the second group disappear, "as though the
afferent paths were competing for control over the cell" (Wiesel and Hubel
1965, 1038).

It took a few more decades for neuroscientists to visualize this synaptic
competition at the level of individual neurons. Antonini and Stryker
(1993) showed that, within 6–7 days after monocular occlusion, the
geniculocortical axons associated with the deprived eye are shorter, and

have fewer branches, than those associated with the nondeprived eye. They could practically watch the outcompeted axons withdraw in real time.

Incidentally, the formation of abnormal ocular dominance columns provides a good example of how constructionist and selectionist mechanisms can work together. Not only do we see the retraction of axons associated with the deprived eye but also the branching and lengthening of axons associated with the nondeprived eye (Antonini and Stryker 1993, 3572). This represents the kind of activity-dependent growth that neural constructionists talk about.

Synaptic competition drives the formation of abnormal ocular dominance columns. What about normal ocular dominance? This remains controversial. In the 1970s, the neuroscientist Pasko Rakic (1976) argued that synapse selection plays a crucial role in normal ocular dominance formation, too. He used staining to show that in the Rhesus monkey fetus, geniculocortical cells are spread diffusely across the visual cortex. There are no well-defined ocular dominance columns. In the second half of gestation, this profusion of synapses has been substantially reduced and the synapses segregate into groups, probably because of spontaneous retinal activity. This supports the profusion-and-elimination model that selectionists endorse (see Katz and Shatz 1996 for discussion). LeVay et al. (1978) found comparable results for the kitten's visual cortex, although, unlike in the rhesus monkey, the segregation of geniculocortical neurons occurs after birth.

However, recent research has called into question the idea that synapse selection is needed for normal ocular dominance columns. Crowley and Katz (1999; 2000) showed that, in the ferret's visual system, the segregation of geniculocortical neurons into ocular dominance columns doesn't require visual activity. They blinded newborn ferrets and showed that segregation still happens normally. This led them to think that normal ocular dominance column formation probably involves some mix of activity-independent mechanisms (such as genetic cues) and activity-dependent elaboration of new synapses but no selection. They also said that earlier results on monkeys and cats that suggested a profusion-and-elimination model were an artifact of bad staining techniques.

Still, we shouldn't be quick to dismiss the role of competitive mechanisms in the formation of normal ocular dominance columns (Price et al. 2011, 210). First, the segregation of axons into ocular dominance columns might depend on spontaneous activity in the thalamus itself rather than the retina. Therefore, Crowley and Katz can't exclude all activity-dependent selection, even in blind ferrets. Second, there's an important distinction between microscopic exuberance and macroscopic exuberance (Innocenti and Price 2005, 959). Macroscopic exuberance points to the role of synapse selection in making connections between different regions of the brain (like the thalamus and cortex). Microscopic exuberance points to the role of synapse selection within a much smaller brain region. Even if Crowley and Katz's work shows that synapse selection doesn't create the macroscopic pattern of connections between thalamus and cortex, it might be involved in the local fine-tuning of those connections.

How far does synapse selection extend? How common is it in the brain? I don't know, but recent work suggests that it's vital for normal cognitive development, so it would be a mistake to try to marginalize it. Some of the most exciting new work suggests that abnormalities in synaptic pruning are involved in cognitive problems like autism and schizophrenia (Chung et al. 2015; although see Feinberg 1982/83 for an early precursor). Additionally, some researchers think the failure of early synaptic pruning is responsible for synesthesia, where a sense impression typically created by stimulating one area of the body gets triggered by another (like a visual impression being triggered by a musical note). Their idea is that because of too little synaptic pruning during infancy, inappropriate cross-modal connections last longer than they're supposed to (Spector and Maurer 2009).

Finally, some theorists think synapse selection is involved, quite generally, in damage-induced neural plasticity. Consider how people who are deprived of one sensory modality, such as sight, often develop heightened acuity in others, such as hearing. Somehow, the parts of the sensory cortex that were originally assigned to transmit visual information are recruited for transmitting auditory information. One speculative theory holds that this cross-modal reassignment is due to synapse selection. It is based on the observation that a portion of the cat's cortex, the anterior ectosylvian cortex, receives inputs from both auditory and visual neurons. This suggests that when a kitten is deprived of vision, the auditory neurons "outcompete" the visual neurons for space, leading to heightened auditory ability (Rauschecker 1995). In short, even if we don't know how common neural selection is relative to other mechanisms of synapse formation, we seem to need it for normal cognitive functioning.

5.4   The Function of Selection

Synapse selection is real, and it's implicated in normal cognitive development. But what is its function? Put crudely, why didn't nature just stick with some mix of genetic cues and neural construction? Early theorists, like Changeux and Danchin (1976, 706) thought they had the answer: "economy of genes." Synapse selection makes it unnecessary to genetically encode the instructions for where to place every single synapse. Instead, the genes merely tell the brain to generate a vast number of synapses and then it lets experience delete the ones that don't prove their mettle. It's worth remembering that when Changeux floated this idea, he was targeting Sperry's chemoaffinity hypothesis that said that most synapses are genetically specified. Next to Sperry's model, it's easy to see why Changeux's profusion-and-elimination model was better, since it required far fewer genes to implement.

Maybe Changeux is right that one function of synapse selection is to afford an economy of genes, but that's probably not its only job. After all, there are alternative mechanisms of synapse formation, like neural construction, and they don't require the genetic specification of each and every synapse either. Considerations of gene economy don't entirely explain why neural selection would evolve instead of some blend of neural construction and genetic cues.

Another potential function for neural selection is a quantitative one. It helps to make sure we have just the right number of connections between pre-synaptic and post-synaptic neurons (Purves 1994, 45). Consider the rat's neuromuscular junction in Figure 5.1. At birth, each muscle fiber is connected to several motor neurons; within weeks, a one-to-one pattern emerges. This one-to-one pattern of connections is crucial for fine motor control. Neural selection makes sure there aren't too many neurons for the job.

Here's the problem with the quantitative hypothesis. If the function of synapse selection is to guarantee a quantitative "match" between the number of pre-synaptic and post-synaptic sites, why not just make sure from the beginning that each muscle fiber is only controlled by a single neuron? That is, why not just set it up that way from the start, rather than go through this cumbersome process?

A more plausible function for synapse selection is that it confers an enormous amount of plasticity onto the developing brain (Innocenti and Price 2005, 958). It helps each brain adjust to its peculiar life circumstances, whatever those circumstances might be. A simple example, drawn from the work of neuroscientist J. Z. Young (1964), shows how synapse selection aids plasticity.

Imagine an octopus that must decide when presented with a new sort of object (say, a crab) to attack or to retreat. Fight or flight? One way it could do this, Young reasoned, is by having a random profusion of synapses. Some cause attack, some cause retreat, and when that crab appears, a response is selected by chance. Those synapses that have bad consequences – that is, that cause pain – are inhibited, and this inhibition ultimately leads to structural changes like axon withdrawal. Thus, the random profusion of synapses, and their selective elimination, helps the octopus tailor its behavior to its environment. As Young says: "That learning occurs by elimination of the unused pathways is a hypothesis that has many attractions. It becomes clear that learning is of the nature of a reduction of the initial redundancy of connections" (Young 1964, 285). Changeux clearly echoed Young's view when he said that to learn is to *eliminate the surplus*.

To give a simplified version of Young's (1964, 156) example, suppose there is one neuron, N, that is triggered directly by crabs, a "classification" neuron (see Figure 5.2). Suppose N synapses onto two target neurons, N_{T1} and N_{T2}, forming two synapses, S_1 and S_2. N_{T1} causes attack, and N_{T2} causes retreat. Suppose that S_1 and S_2 are weighted the same, so that there is a 50/50 chance that a given response will happen (Figure 5.2a). This represents a "divergent" pattern of connections.

Now a crab appears, which activates N, which activates N_{T1}, which causes the octopus to attack. A moment later, the octopus experiences the searing pain of a counterattack. The pain receptors control a device that inhibits S_1. This makes sure that, when that same crab appears, or others that look like it, the octopus will probably retreat (Figure 5.2b). Had the crab made a tasty snack instead, the whole configuration would be flipped the other way around. In short, the mechanism of profusion-and-elimination supports neural plasticity. It's not obvious how purely

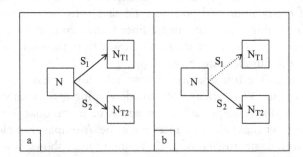

Figure 5.2   A simple example of synapse selection. The first panel (a) shows a neuron N innervating two target neurons, N_{T1} and N_{T2}, yielding two synapses, S_1 and S_2. As a result of experience, S_1 is inhibited, leaving S_2 (as shown in [b]).

constructionist mechanisms could achieve the same flexibility. How would a constructionist device "know" that it ought to form a synapse between N and N_{T_2} rather than N and N_{T_1}? How but for this synaptic game of trial-and-error?

Set aside functions and turn to mechanisms. How does synapse selection happen? This question splits into two. First, what exactly are these synapses competing over? What are they trying to get? Second, what causes their elimination? That is, when a synapse cannot get enough of whatever it's trying to get, what forces it to withdraw? One possibility is that synapses are competing over a limited amount of *neurotrophic factor* provided by the target neuron (e.g., Elliott and Shadbolt 1998). In this picture, neurons are, quite literally, fighting over food. A second possibility is that axons, and even whole neurons, can actively "sabotage" others (Deppmann et al. 2008; Gibson and Ma 2011, 190–191). More recent research has focused on microglia, which travel around and consume unused dendritic spines (Tremblay et al. 2010; Stephan et al. 2012). The point is that we're still waiting for the empirical data to flow in.

In this chapter, I've often described synapse selection by saying that the "useful" connections are kept and the "useless" discarded, but useful to whom? In Young's octopus example, synapse selection is useful to the octopus, but that's not necessary for synapse selection. It could be that one synapse is simply better than another at extracting a common resource it needs to persist, even if its persistence hurts the organism. That would still be a legitimate function-bestowing process. A new function would arise, to be sure, but it'd be a function for the synapse, not for the organism (just like a part of a parasite can have a function for the parasite but not for the host). Even though the standard examples of synapse selection involve some benefit to the organism, I'm not committed to any organism-level bias in how functions are doled out.

5.5 How the Brain Makes Functions

I talked about the function of synapse selection, considered as a general-purpose device for making and keeping synapses, but how does synapse selection create *new* functions? Consider the octopus and the crab. Here, synapse S_2 was selected over S_1, by synapse selection, because it caused the octopus to retreat when crabs were around. In my view, if S_2 was selected by synapse selection because it caused retreat in response to crabs, then its *function* is to cause retreat in response to crabs. (Technically, this is a *response function*, which I'll discuss in Chapters 11 and 12.) I'll defend this idea in the next chapter.

Before synapse selection, S_2 didn't have retreat as its function. After synapse selection, retreat became its function. Before selection, S_2 might have had other functions; I'll discuss this potential overlap of functions in the next chapter. To the extent that it actually caused the octopus to retreat before synapse selection, that was just a lucky accident, not a function – just like the first zebra that ever developed stripes.

Are there other interesting cases of *differential retention* in the natural world apart from synapses and neurons? Here are four other cases that deserve consideration: beliefs, behavior dispositions, clonal organisms and, quite possibly, social institutions. I'll go through each in turn, but keep in mind that this is just a sketch; I don't want to rest my case for GSE on any of them. I want to show the power of differential retention in the world and, corresponding to that, indicate the breadth of application of my view of function.

First, consider beliefs. As discussed in Chapter 4, Papineau (1984, 557) proposes that beliefs acquire novel functions through competitive interactions. He describes beliefs in terms of their causal roles – that is, in terms of their inputs, outputs, and interrelations. He points out that these psychological entities can be reinforced over others because of their consequences: "new concepts are 'fixed' by learning precisely when the beliefs they give rise to are advantageous." That sounds like a plausible picture, but beliefs, Kingsbury reminds us, don't reproduce; they merely stick around better or worse than others. They cannot acquire functions in the standard way, via differential reproduction or copying. If we drop the requirement that functions need reproduction, we can give new functions to them. This strikes me as a benefit. Clearly, I might believe something false (for example, that I am the object of a secret admirer) over something true, because it inflates my ego or helps me cope with a sense of personal worthlessness. That's a function without reproduction.

For a second example, we should reconsider the selection of behaviors by operant conditioning. Millikan thinks such behaviors have functions because they undergo differential reproduction or copying. Is this really a case of "reproduction?" That seems like stretching the very concept of reproduction (Grantham 2001; Kingsbury 2008, 502). Is an instance of behavior at a given time (say, my sipping coffee on a Tuesday morning) a reproduction of an earlier instance of that behavior (my sipping coffee on Monday)? Reproduction involves a tight cause-and-effect coupling between parent and offspring; think of the production of daughter cells by mitosis. In operant conditioning, however, there is a behavioral disposition (the disposition to sip coffee) that produces a group of behavior tokens that resemble one another. The first token doesn't cause the second – or if it does, it does so in an extremely circuitous way. It's more natural to describe this situation in terms of the differential *retention* of behavioral *dispositions*, rather than the differential *reproduction* of behavior *tokens*. Again, the traditional selected effects theory doesn't let us give (direct proper) functions to learned behavioral dispositions. The theory I'll present in the next chapter does.

Third, Frédéric Bouchard (2008) has argued that sometimes natural selection merely involves the differential persistence of things rather than their differential reproduction (also see Doolittle 2014; 2017). A good example comes from clonal organisms. The world's largest living organism, Pando, is an aspen grove in Utah's Fishlake National Forest. Biologists consider Pando to be a single organism on account of its massive underground root network and its nearly genetically identical trunks. Although Pando is capable of sexual reproduction, it seems to have done better for itself, and outcompeted conifers for space simply by growing.

Now suppose that, through a gene mutation, a small part of this massive organism acquires a new trait, one that helps it expand significantly over other parts of Pando. Should we deny a function to it because it's not, technically speaking, helping Pando reproduce? I don't see why. For example, most of Pando's trunks are over a hundred years old because cattle and deer eat the new saplings (Rogers and St. Clair 2016). Suppose some of its trunks develop a gene mutation that changes the taste of saplings and thereby deters cattle so that trunks with the mutation become more frequent. That would be a function by virtue of differential persistence, not reproduction. As I'll note in the next chapter, *clade selection* might also be an example of differential retention without differential reproduction (Doolittle 2017).

Note, incidentally, that this hypothetical example would be an instance where differential retention gives rise to a new direct proper function that isn't, at the same time, a derived proper function of anything. If you remember, when a synapse is retained by synapse selection because of some effect, it comes to have a direct proper function (since it was selected for that effect) and it also comes to have a derived proper function (since it was produced in the right way by a general-purpose mechanism that carries out its function by producing synapses like that). In the case of antibody selection and neural selection, this convergence, or overlay, of direct and derived proper functions is the norm. The hypothetical example

I described above – where one part of Pando undergoes rapid expansion due to a mutation that deters browsing – would be a direct proper function (since there's differential retention) but not a derived proper function (since there's no mechanism that has the function of producing mutations like those). The two can, in theory, come apart, even though they usually hang together.

Finally, it's worth considering whether *social institutions* or their parts undergo anything like differential retention. Here, I am indebted to Reid Blackman, who has been using the selected effects theory to work out a theory of the functions of social roles and institutions. Suppose, to use one of Blackman's examples, we are interested in the function of judicial punishment. Is it deterrence, retribution, oppression of the poor, or something else? Here's one way to think about it. Suppose that, at some point in the past, a society developed various mechanisms for deterring violent crime. Suppose these included judicial punishment (e.g., prison, the imposition of physical pain, or death), exile, and reconciliation. Suppose that the first of these managed to deter wrongdoers better than the latter two, and that is why it was retained. Then we can say that the function of judicial punishment is deterrence, because that caused the differential persistence of that institution. Again, this is just a sketch, not a full-fledged theory. The point is to recognize the power of differential persistence to shape nature and society.

CHAPTER 6

A Generalized Selected Effects Theory of Function

This chapter presents and defends the generalized selected effects theory (GSE) of function. One benefit of this view is that it accommodates neural selection and natural selection, too, but that's not the main argument for holding it. The main argument for GSE relies on parity of reasoning. Anyone who accepts the selected effects theory should accept GSE instead on pain of inconsistency.

Section 6.1 will set out the theory itself, along with the main argument for GSE. Section 6.2 sets out six arguments against GSE, in order of increasing difficulty. In responding to some of these arguments, I invoke the notion of a *population*; Section 6.3 explicates this idea in more detail. Equipped with the notion of a population, we're prepared to confront a seventh argument against GSE that is tricky enough to merit its own section (Section 6.4).

6.1 The Theory

Recall that on the traditional selected effects theory, functions are given, in the first place, to things that reproduce. In Millikan's (1984) terminology, functions belong to the members of "reproductively established families." In Neander and Rosenberg's (2012, 618) more recent parlance, they belong to members of a "lineage of traits parsed by changes in the selection pressures operating on it." The traditional theory can also give functions to antibodies and, somewhat problematically, to behaviors. Even this liberal reading of the selected effects theory, however, contains an arbitrary restriction. Why do functions need reproduction at all?

Instead, here is how we should think about functions:

> The generalized selected effects theory (GSE): A function of a trait is an activity that led to its differential reproduction, or its differential retention, in a population.

93

The first part, differential reproduction, captures the exact same entities the traditional theory does: organisms, antibodies, and perhaps (again, problematically) behaviors. The second part, differential retention, captures neural structures, behavioral dispositions, and desires. Perhaps it can give functions to some clonal organisms and social institutions, too, but fleshing that out in a rigorous way is a project for the future. I will explain the third part, "in a population," momentarily.

The core argument for GSE is simply a parity of reasoning argument. Consider why we accepted the traditional selected effects view. We did so because it made sense of the three big puzzles of function: the function/accident distinction, the possibility of dysfunction, and function's explanatory depth. Since GSE solves all the same problems, minus an arbitrary restriction, we should accept it.

I'll use J. Z. Young's simple octopus example (Figure 5.2) to show how GSE solves all three puzzles. If you recall, one synapse (S_2) was retained over another (S_1), because it caused a useful effect – it helped the octopus escape from crabs. GSE says that, after synapse selection, a function of S_2 is to cause retreat from crabs, since that explains why S_2 was retained over S_1.

Now take those three puzzles in turn. The first is the distinction between functions and accidents. The reason S_2's function is to cause retreat is because that's what it was selected for. It might have other benefits, too, but those are lucky accidents, not functions. In the summer, some rock crabs have high levels of an algae-caused neurotoxin, domoic acid; it can be deadly in large quantities. Suppose that, by avoiding crabs, an octopus avoids shellfish poisoning, too. Thus, S_2 has at least two benefits: It helps the octopus avoid pain and it helps the octopus avoid poison. However, only the first is a function, and not the second, since the synapse wasn't selected for avoiding poisoning. Avoiding domoic acid poisoning doesn't explain why it's there.

GSE also captures the possibility of dysfunction. I'll develop this point in Chapter 8, but the main idea is simple. Suppose S_2 acquires the function of causing retreat, since that's what it was selected for by synapse selection. Suppose that, some time later, it can't cause retreat anymore because of an intrinsic impediment, such as the loss of a myelin sheath. In that case, S_2 has a function it cannot perform, and therefore (give or take some niceties to be discussed in Chapter 8) it's dysfunctional. GSE gives us normativity without natural selection between organisms. Neural selection is quite sufficient.

Finally, GSE makes sense of function's explanatory depth. Suppose a neuroscientist, studying the visual system of that octopus, asks, "Why is S_2

there?" In other words, suppose one wishes to know why N synapses onto N_{T2} rather than something else (N_{T1}, or both). One correct answer is: "S_2 is there because it causes the octopus to retreat from crabs." Retreat is the effect that caused the perpetuation of S_2 over S_1.

Other philosophers have developed views that are similar to mine in some respects. I'll take the rest of the section to map out points of convergence and divergence. First, GSE owes a debt to thinkers like Wimsatt (1972), Millikan (1984), Papineau (1984; 1987; 1993), Godfrey-Smith (1992) and Griffiths (1993) who were willing to extend the traditional selected effects theory outside of natural selection between individuals. The difference between my view and theirs is that they still restricted functions to reproducing things (with the possible exception of Wimsatt, who didn't endorse the selected effects theory anyway). As I argued in Chapter 4, this unfortunate restriction made it impossible to extend the theory in a consistent way.

Recently, Bouchard (2013) developed a theory of function that, like mine, gives functions to things that undergo "differential persistence" rather than differential reproduction. His view and mine share a spiritual affinity, although we developed our ideas independently (see Garson 2006 for an early attempt). Earlier, Bouchard (2008) argued that natural selection merely requires differential persistence, not reproduction. He was mainly concerned with identifying what counts as evolutionary "success" for colonial organisms like corals, and clonal organisms like aspen groves, that thrive by something like differential growth, not only by breeding. At the time, however, he hadn't applied the lesson to the functions debate.

Nonetheless, Bouchard's theory differs from mine in three main ways. First, his is an ahistorical, forward-looking theory. He rejects etiological accounts because he wants to give functions to things on their very first appearance rather than after a few rounds of selection (Bouchard 2013, 94). Second, my view emphasizes the notion of a "population" in order to avoid some liberality problems; Bouchard doesn't do so explicitly. Third, he was mainly thinking about ecology and evolution when he developed his theory; I was mainly thinking about neuroscience and mind. I'll come back to his view briefly in Chapter 9.

The biologist W. Ford Doolittle (2014; 2017), like Bouchard, argues that natural selection only needs differential persistence, not differential reproduction. Doolittle has written on the topic of *clade* selection: that is, the question of whether there can be selection between whole lineages of organisms rather than individuals. He thinks that there can be, but only if we extend evolutionary natural selection to include differential persistence in addition to differential reproduction. As he puts it: *"differential persist-ence alone can support a process that we cannot dismiss as a form of"* [evolutionary natural selection] (2017, 280; emphasis his). Since Doolittle was not writing about functions, his thesis is somewhat orthogonal to mine. I don't have an official stance on defining "natural selection." The point is that Doolittle recognizes the power of differential persistence to sculpt nature in meaningful ways.

Finally, Lewens (2004, 129) hints at a theory like GSE. He points out, rightly, that the main rationale for the traditional selected effects theory is that it solves the big three puzzles of function. The problem, he says, is that any mere sorting process can do the same. For Lewens, this is actually a vice for the theory, not a virtue. I'll return to his objection at the end of the next section.

6.2   Six Problems for GSE

Here are six problems for GSE, in order of increasing difficulty. Showing how GSE handles objections not only fortifies the view but also helps to clarify nuances that might go unnoticed. The first three problems are about synapse selection. They question whether synapse selection, per se, should count as a function-bestowing process. The last three affect GSE itself, quite independent of synapse selection.

1 *The vacuity objection.* In a largely negative review of Edelman's book, *Neural Darwinism*, Francis Crick (1989, 247) urges us to stop talking about synapse selection: "to avoid confusion I feel the term 'selection' is better not used. Moreover, almost everybody's theory could be called a theory of synaptic selection." If I understand him correctly, his complaint is that on almost anyone's theory, we start out with a vast collection of synapses, some of which are created or strengthened by experience, and some of which are weakened or eliminated. Isn't that a kind of selection process? His observation seems to render the whole idea of synapse selection vacuous.

If Crick were right, that would be devastating for my theory of function. For it would mean that virtually any activity-dependent synapse change would count as "selection," and hence could create new functions. Panic attacks would create new functions, simply because they strengthen some synapses and not others; so would grand mal seizures, according to the traditional (albeit controversial) "kindling" theory.

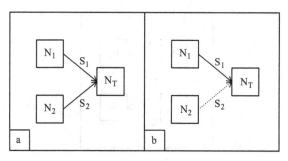

Figure 6.1   A simple example of synapse selection. Panel (a) shows two different neurons, N_1 and N_2, innervating the same target N_T, and yielding two synapses, S_1 and S_2. After synapse selection, only S_1 remains (panel [b]).

Fortunately, not everyone's theory is a selectionist theory. In order for synapse selection to happen, it's not enough that some synapse, somewhere, is strengthened, and some synapse, somewhere else, is weakened. By the same token, for natural selection to happen it is not enough that some animal, somewhere, lives, and another one, somewhere, dies. Rather, the differential retention of synapses must occur in the same population. I will flesh this idea out more shortly, but the rough idea is that there must be persistence-relevant interactions between the synapses in the group. In synapse selection, these interactions are competitive. One synapse is eliminated because another is retained. In neural constructivism, nothing like a competition need take place.

Here's a simple example. Figure 6.1 illustrates a case of synapse selection. At time t1, there are two neurons, N_1 and N_2, which innervate the same target, yielding synapses S_1 and S_2 (panel a). At time t2, as a result of competitive interactions, S_2 is eliminated and S_1 is retained (panel b). According to GSE, at t2, S_1 receives a new function: namely, whatever it did that caused its differential retention. (This figure is similar to Figure 5.2, except that it illustrates a convergent pattern of innervation rather than a divergent pattern. Nothing important hangs on that distinction here.)

Figure 6.2, in contrast, shows a case of normal constructionist growth. There are two synapses, S_1 and S_2, but they don't innervate the same target. N_1 activates N_2, so S_1 is strengthened. N_3 fails to activate N_4, so S_2 is weakened (one can think of this as disuse-related atrophy). The crucial bit is that although both cases illustrate something like "differential retention," only the first involves synapse selection, because only the first

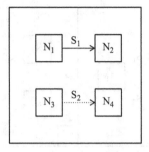

Figure 6.2   An example of normal constructionist growth. N_1 innervates N_2, yielding S_1; N_3 innervates N_4, yielding S_2. S_2 atrophies as a result of disuse. There is a kind of differential retention but it is not synapse selection and hence does not create a new function.

involves a persistence-relevant interaction between the synapses. Only in the first is there a genuine population of synapses (see Section 6.3).

2 *The arbitrariness objection.* Here is a second criticism. GSE implies that synapse selection creates new functions but that synapse construction doesn't. This might strike some people as arbitrary. After all, both synapse selection and synapse construction are activity-dependent mechanisms for the formation and maintenance of new synapses. Why does synapse selection create new functions but not synapse construction?

Consider Hebb's rule: "Neurons that fire together, wire together." Suppose that, due to the joint activity of N_1 and N_2, a synapse S is formed between them that helps to ensure their continued joint activity. We can even suppose that S benefits the organism. Maybe it is even a matter of life or death. GSE implies that this activity-dependent strengthening of synapses does not, all by itself, make new functions, since there's no selection involved.

The argument is actually quite similar to an argument that's been raised against the traditional selected effects theory. The argument is this: There are a lot of processes that lead to the evolution of useful traits. Natural selection is only one of them. In theory, a useful trait can evolve by genetic drift. It can evolve as a byproduct of something else undergoing selection. Why restrict functions to those traits that evolved by natural selection? What's so special about natural selection? Sometimes this argument is packaged together with the subtle (or not-so-subtle) insinuation that selected effects theorists don't know much about biology or, even worse, that they're a bunch of closet adaptationists.

By now, this argument should have lost its grip on us. The reason natural selection creates new functions, but genetic drift doesn't, is that natural selection lets us make sense of the explanatory depth of functions: that is, how function attributions can double as causal explanations. If the zebra's stripes were selected for deterring flies, then zebras have stripes because stripes deter flies. The same cannot be said if the stripes came from drift, or if they were an inevitable byproduct of some other trait. Any benefit they yield would be a lucky accident.

The same lesson can be applied in the neural context, too. If synapse S_1 is selected over S_2 because it does E, then S_1 is there (rather than S_2) because it does E. Suppose, however, that S_1 is there purely randomly but has some useful effect E. Then it's not true to say S_1 is there because of E. Therefore, not just any mechanism of synapse formation can create new functions. While it may seem arbitrary at first to let neural selection create new functions but deny that privilege to construction, a little reflection on first principles should dispel any doubts on the matter.

Here is one way to think about the difference between selection and construction in the neural context. The main purpose of neural construction is to strengthen, amplify and reinforce old functions, rather than create new ones. It's like turning up the volume on a radio rather than changing the station. Or, as I suggested earlier, it's like dredging a river rather than forming a new branch.

3 *The redundancy objection.* Here is a third criticism. One might think that synapse selection shouldn't create new functions, because any new functions it creates would be redundant. After all, neurons and their components already have functions because of natural selection between organisms. They were designed to transmit and process information in the brain. Natural selection already gave them functions and they don't need to double up on them.

I agree that neural structures already have functions from birth onward because of natural selection. Consider synapse S_2 in our octopus (Figure 5.2). Before synapse selection, the synapse, as a synapse, already had the function of transmitting chemical signals. This was a generic function it shared with most other synapses. Natural selection gave it that function. That doesn't mean any new functions it acquires through experience are redundant or unneeded. A trait can have more than one function. After synapse selection, it acquired a new function: namely, to facilitate retreat in response to crabs. A trait can have multiple functions since it can be the outcome of multiple selection processes.

At any rate, the traditional selected effects theory already lets a trait have multiple functions if it's involved in multiple selection processes. That's because the theory recognizes different kinds of selection processes. A trait can have one function by natural selection acting between individuals and another by natural selection acting between groups. At the extreme, there can even be conflicts between levels of selection and hence tensions among the functions of one and the same trait; the tension between aggression and docility in water striders is a great example. I see this as an intriguing consequence of the theory and not an embarrassment. It will be particularly important when we come to the topic of mental illness that in theory might exhibit such conflicts (see Chapter 11).

4 *The argument from derived proper functions.* Now we're moving into criticisms of GSE generally, irrespective of synapse selection. There is a kind of redundancy problem that stems from reflecting on Millikan's distinction between direct and derived proper functions. Here is a plausible line of thought: "Granted, on the traditional selected effects theory, neural selection cannot make direct proper functions, but it can make derived proper functions. Consequently, nobody needs GSE. Millikan's theory already lets us give functions to everything you wish to give functions to." In other words, why accept GSE, when we can just accept the traditional selected effects theory and supplement it with the distinction between direct and derived proper functions?

It's helpful to recall how Millkan's distinction works. In Millikan's view, in order for a trait to have a direct proper function, it has to be the kind of thing that reproduces (or "gets copied"). It must create parent–offspring lineages. Parts of organisms reproduce (or "get copied") by ordinary selection between individuals. Behaviors reproduce, in her opinion, by operant conditioning, but since neurons, synapses, and neural groups don't reproduce, they don't make new direct proper functions. Neural selection can, however, make new functions of a sort. It makes derived proper functions. Neural selection, as a general-purpose brain mechanism, was shaped (let us assume) by natural selection, because it helps creatures navigate their local environments. Any novel configurations of synapses that neural selection creates, in the course of discharging its function, have derived proper functions. The function of the producer slides seamlessly over to the produced, without needing an extra round of selection.

I agree that, if all we care about is "getting the job done," then the traditional selected effects theory, along with the distinction between direct and derived proper functions, will do that job. If all we want is a theory that lets us give functions, somehow or another, to the products of synapse selection (say), then that mix of theories will work. Philosophers, however, usually care about more than just "getting the job done," and sometimes not even that. They want a theory that gets the job done without arbitrary restrictions or ad hoc extensions. Since my theory gets the job done without any arbitrary restrictions, it's a better theory.

Another way of putting the point is that Millikan's approach lacks a certain elegance one might wish for in a theory of function. It first banishes nonreproducing entities from the realm of direct proper functions. Then it helps itself to another distinction, between direct and derived functions, to confer functions of a lesser rank upon them. But why did we expel them in the first place? And why rely on these cumbersome tactics to bring them back in? I have no problem with direct and derived proper functions, per se; it's a tolerably well-defined distinction (though some disagree – see Preston 1998), but we shouldn't have to rely on this distinction to do the work that our core theory should do all alone.

5 *Against parity.* Here is a fifth potential objection, this time aimed at parity of reasoning. My main argument for GSE is that it solves all the puzzles of function, without pointless restrictions. Parity of reasoning demands that we accept it. Maybe there are other reasons, however, to accept the traditional selected effects theory over GSE, quite independent of its ability to solve those three puzzles. If so, then parity of reasoning wouldn't be enough.

What would be an argument for the traditional selected effects theory that doesn't apply to GSE? One might argue that we should prefer the traditional theory because it better captures how biologists actually talk. After all (the argument might go), sometimes, when biologists give functions to traits, all they mean is that the trait evolved by natural selection for some end (see Lorenz 2002 [1963], 11; Maynard-Smith 1990, 66). The traditional selected effects theory matches real biological usage and GSE doesn't.

This is an interesting argument, but I think it fails. After all, many philosophers reject the selected effects theory because they do not think it reflects explicit biological usage very well (e.g., Schlosser 1998, 304; Wouters 2013, 480). It would be a little strange to hear someone insist that we accept it on the ground that it does reflect it.

It might seem as if I'm backtracking on something I said much earlier (Chapter 1). I said that a good theory of function should be sensitive to how biologists actually use the term "function." Specifically, I said the best way to figure out what functions are is to look carefully at the way biologists use the term, and then see if there is anything in the world that matches their usage. Now I argue that we shouldn't care too much about how biologists use the term "function," and that my preferred theory doesn't match their usage anyway.

There's no real contradiction. My view is that, sometimes, when biologists give functions to traits, they are implicitly committed to GSE, even if they are not explicit about it. It's not as if biologists have something like GSE in mind when they give functions to traits. When biologists give functions to traits, they often use "function," explicitly, in a way that respects the distinction between function and accident, function's normativity, and its explanatory depth. Selection is the only thing in the world that can underwrite all of those features of ordinary biological usage without producing thorny liberality problems. As a consequence, biologists are implicitly committed to something like GSE, even if they don't know it yet.

At any rate, most proponents of the traditional selected effects theory don't argue that we should accept it because it matches explicit biological usage. Instead, they say we should accept it because it solves one or more of the three puzzles of function. Wright (1973, 159) defends his precursor to the selected effects theory by pointing out that it gives explanatory depth to functions. Millikan (1989a, 296) and Neander (1991, 180) defend it by saying that it explains the possibility of dysfunction. Lewens (2004, 129) points out, as I do, that it satisfies all three puzzles of function, although he tries to use this fact in an argument against the selected effects theory – more about this soon. I don't know of any good arguments for the traditional selected effects theory other than its ability to solve the three puzzles, but GSE solves those puzzles better than the traditional theory. Parity of reasoning still works.

6 *The liberality objection.* In Chapter 3, I rejected all feedback theories on the grounds that they're too liberal. One might level the same charge against GSE. To tweak an example of Kingsbury's (2008, 496) somewhat, imagine a bunch of rocks scattered along a beach. Some rocks are harder and some are softer. The softer rocks tend to erode more rapidly, leaving the harder behind. This is a kind of differential persistence. Does GSE force us to say that hardness is a function of rocks?

Lewens (2004, 128) raises a similar problem, but this time for the traditional selected effects theory. The basic rationale for that theory, he notes, is that it solves the three puzzles of function. Then he says that any mere sorting process can do the same. In a sorting process, there is a collection of entities, and there is some rule or principle that causes some of them to be sifted out: "There is variation across a collection of items, and differential propensities among the items to survive some kind of test, but no reproduction" (p. 127). Witness the phenomenon of longshore drift, which causes large rocks to accumulate at one end of a beach and small ones at another. His entire argument, if I understand it correctly, is an attempt at a clever *reductio ad absurdum* of the traditional selected effects theory. If you accept the theory, you have to accept (by parity of reasoning) that mere sorting creates new functions. If you accept that, then you have to accept that longshore drift creates new functions – which is absurd.

As a side note, Lewens doesn't actually reject the traditional selected effects theory. He says that it is a "waste of time" to try to distinguish genuine functions (like those of the zebra's stripes) from "as-if" functions (like the effects of longshore drift). He says we should take a "deflationary" attitude about functions (p. 18). The attitude he recommends is not so much outright rejection but humble recognition of its shortcomings.

Lewens's pessimism is far too premature. GSE doesn't make us give functions to rocks on account of their hardness or size. Nor must we let longshore drift make new functions. GSE only says that a trait's activity is its function if it led to differential retention or differential reproduction within a population. A bunch of rocks scattered on a beach, or heaped up into piles, doesn't count as a population, so GSE excludes it.

6.3 Functions and Populations

What is a population? And, more importantly, what justifies this restriction? Unfortunately, only a handful of philosophers have tackled, directly, the question of what populations are; see Godfrey-Smith (2009), Millstein (2009), and Matthewson (2015) for notable attempts. An idea woven throughout the sparse literature is that in order for a collection of things to form a population it's not enough that they're close to one another in space, or even that they have a shared history. Rather, they have to interact in the right way. As Sober and Wilson (1998, 93) put the point in talking about group selection: "Individuals belong to the same group because of their interactions, not because they are elbow-to-elbow."

What sorts of interactions create populations? The emerging consensus is that populations require *fitness-relevant* interactions (Godfrey-Smith

2009, 52; Millstein 2009, 271; Matthewson 2015, 180). Recursively put, for A to belong to the same population as B, A must affect B's fitness, or A must affect the fitness of some entity C which is part of the same population as B. I'll clarify the details of this definition shortly.

Even with this very minimal characterization at hand, a bunch of rocks on a beach isn't a population. There are no fitness-relevant interactions between them. (To be precise, since rocks don't have fitness scores, we should say there are no persistence-relevant interactions between them.) The rate at which one rock erodes doesn't affect the rate at which others erode. Even before we delve into the messy details of what else must be true of populations, we know that rocks on a beach don't count. We don't need to solve any subtle controversies about what populations are to get the right answer.

What's the justification for restricting functions to things inside a population? Is this just another one of those ad hoc devices for dodging counterexamples? Or does it flow out of a deep principle? Restricting functions to the members of populations is not an ad hoc device, but it emerges by reflecting on what selection is. For selection always involves a population. Consider a group of rats in Manhattan's Central Park, and another group of rats in San Francisco's Golden Gate Park. Suppose the Central Park rats have a higher fitness than the Golden Gate Park rats because of some behavioral difference. Is this natural selection? Intuitively, it isn't, since the groups aren't parts of the same population. They might as well be on different planets as far as natural selection cares (Godfrey-Smith 2009, 48). There has to be something that ties individuals into a bundle that selection then acts on. There's nothing ad hoc about saying that selection always involves a population. Now, we can recognize that natural selection always involves populations without being committed to any particular construal of what populations are. The further claim I'm making here is that fitness-relevant interactions constitute populations.

There's more that should be said about what populations are. Do any fitness-relevant interactions count – cooperative or competitive ones – or only the competitive ones? Godfrey-Smith (2009, 52) says that competitive interactions are especially important for Darwinian populations: "competition is an especially Darwinian glue." For my own purposes, it does not matter whether we include both kinds of fitness-relevant interactions or only competitive ones. (Incidentally, synapses can cooperate as well as compete; a group can mutually reinforce each other in a phenomenon known as "associative plasticity" – see Price et al. 2011, 304.)

As I noted above, the way I deploy the concept of a population differs in one crucial respect from how other philosophers use it. In my view, in order for a group of entities to count as a population, its members need not reproduce. It's enough that they persist for various lengths of time, and that they affect each others' persistence. If we're willing to expand the concept of a function-creating selection process to include selection between differentially persisting entities, we have to be ready to tweak a bunch of other concepts, too. For example, once we were willing to admit that ontogenic selection processes can create new functions, we had to suitably amend the meanings of "normal environment," "recent history," and so on. That's all I'm doing when I expand the idea of fitness-relevant interactions to include persistence-relevant interactions.

David Papineau pointed out to me that there's another way to solve the problem of rocks on a beach, one that doesn't invoke interactions at all. We might say that the differential persistence of rocks on a beach doesn't create new functions because the selection process at hand can't give rise to complex adaptations. That is because there's nothing like *cumulative selection* happening. Consider the evolution of the human eye, where adaptations build on adaptations to craft highly complex organs. In the rocks on the beach example, there is no cumulative selection, hence no complex adaptations – and no functions, either (see Garson and Papineau in prep).

Doolittle (2017) suggests a similar strategy in his discussion of natural selection. As I noted, he thinks of evolutionary natural selection broadly enough to include the differential persistence of clades, and not just differential reproduction. He worries, however, about whether differential persistence will give rise to counterexamples. He raises Okasha's (2006, 214) concern that selection between differentially persisting individuals "is not very interesting, and will not lead to adaptation." In response, Doolittle points out that even though clades don't reproduce, they do exhibit something like cumulative selection: "within a growing clade each subclade and species lineage is potentially a platform for further innovation (Doolittle 2017, 288)." Of course, Doolittle was not talking about functions, but the point could carry over. He might want to restrict functions to entities that undergo both differential persistence and cumulative selection but drop the requirement for interaction.

This is a reasonable strategy, but it goes too far by excluding too much. Consider the evolution of dark wing coloration in the famous peppered moth, *Biston betularia* (Lewontin 1998). In the 1840s, almost all of the peppered moths around Manchester, England, were light colored. As a

result of industrial smog, the trees they rested on became dark; by the
1890s, the vast majority of peppered moths acquired dark coloration. This
strikes me as a paradigm function-bestowing process. Although dark wing
coloration came to have the function of camouflage over a period of
decades, there was nothing like cumulative selection taking place. One
variety just out-reproduced another. One might quibble with the details of
the example, but my point is that cumulative selection shouldn't be a
condition for new functions.

6.4   A Harder Liberality Problem

Here is a new counterexample, that is tougher to tackle. Suppose we
change the rocks example. (I thank Karen Neander for this objection.)
Imagine a group of rocks piled up on top of one another. Whenever the
waves crash in, they jostle each other. The harder rocks not only with-
stand erosion better than the softer ones, but they contribute to the
erosion of the softer ones. Now, there is differential retention of rocks as
well as persistence-relevant interactions between them. Even so, it runs
against both intuition and ordinary biological usage to give functions
to rocks.

We can avoid the force of this counterexample by reflecting more deeply
on what populations are. In a recent paper, Matthewson (2015) argues that
fitness-relevant interactions aren't enough for making populations,
although they're necessary. Here is one way of showing why they're not
enough (loosely based on his example on page 183). Suppose there are two
groups of rats that live on either side of a stream. Suppose that the two
groups are isolated from each other because of the stream, but the rats in
each group interact among themselves. Suppose that occasionally, a rat
from one group manages to find its way across the stream, where the rats
on the other side promptly kill it. According to our recursive definition,
these two groups make a single population, but that seems wrong. There
are two clear-cut populations, not one, despite the occasional migration.
How do we fix this?

Matthewson argues that for a bunch of creatures to form a single
population, not only must there be fitness-relevant interactions between
them, but there also must be a high degree of "linkage." Roughly, this
means that, on average, each member of the population must have fitness-
relevant interactions with several other members of their group – not just a
few. More specifically, the linkage within a group is the ratio of the actual
number of fitness-relevant interactions in that group to the total possible

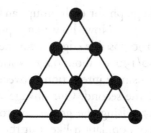

Figure 6.3   A pile of rocks represented as an undirected graph.

number of fitness-relevant interactions. The closer that ratio is to 1, the more population-like the collection is.

We can characterize linkage more precisely by using a graph. Suppose we represent a group of individuals as an undirected graph (as in Figure 6.3). Each node in the graph represents an individual and each edge represents a fitness-relevant interaction. In other words, we draw an edge between two nodes if the individuals they represent have had at least one fitness-relevant interaction over their lives. This gives us the actual number of edges. Then we find the total possible number of edges. We do that by counting how many edges there would be if each node were connected by an edge to every other node. (Assume there are no loops, since individuals can't have fitness-relevant interactions with themselves.) If there are n individuals in a population, the total possible number of edges, excluding loops, is $(n(n-1))/2$. For example, if there are 10 individuals, and each bears fitness-relevant interactions with every other, then there are $(10 \times 9)/2 = 45$ edges.

With this basic picture in mind, there are at least two ways we can apply it for deciding whether a collection of individuals counts as a population. The first is to apply a threshold, and insist that a collection only counts as a population if it has a linkage of, say, 0.5. The problem is that the selection of any specific threshold seems a bit random. Why not 0.4, or 0.3, or 0.9 instead? A second way is to treat the idea of a population as a graded concept. (I thank John Matthewson for helping me think about these options.) We can say that, the closer the linkage score is to 1, the more population-like the collection is. The graded solution is preferable since it isn't arbitrary.

Now consider our two groups of rats separated by a stream. Each group has ten rats. Suppose, in each group, the linkage is perfect: that is, each member interacts with each other. (There are 45 edges in the corresponding graph.) Now consider the linkage of our gerrymandered group. Since there are 45 edges in the graph for each group, and there are two groups, there are a total of 90 edges. Now, how many possible edges are there? There are $(20 \times 19)/2 = 190$ possible edges. That means the total linkage in this entire collection is 90/190, or about 0.47. Notice that this score would hardly change if we add a single interaction between a member of the first group and a member of the second. Our linkage would then be 91/190, which is still about 0.47. Using the graded approach, each group, considered on its own, is very population-like, but the entire collection is not very population-like.

After this somewhat long journey, we can finally explain why a pile of rocks is not very population-like. This is because the pile has a low degree of linkage. Each rock in the group, at best, only affects those rocks that are immediately adjacent to it but has a negligible impact on the others. To clarify, suppose there are ten rocks. Suppose they are piled up on one another in a pyramid shape, as in Figure 6.3. Suppose each rock only has persistence-relevant interactions with its immediate neighbors. Then the linkage of the group is $18/45 = 0.4$. According to the graded approach, the collection is not very population-like.

Matthewson pointed out an interesting consequence of this way of doing things. Let's agree that being a population is a matter of degrees. If that's correct, then functions come in degrees, too. What the linkage analysis shows is that the rocks in our pile have, at best, a very low degree of functionality, and much lower than the level of functionality we typically associate with the products of neural selection or natural selection.

Consider, in contrast, a typical case of neural selection. Think about the rat's neuromuscular junction (Figure 5.1). At birth, multiple motor neurons synapse onto the same muscle fiber. At the end of the selection process, only one remains. Therefore, the retention of any one synapse entails the elimination of all the rest. The linkage in our population of synapses is 1, or very close to 1, quite unlike our pile of rocks.

I realize that more can be said about what populations are; the existing literature is just scratching the surface of a complex and interesting topic. However, I hope that I've said enough to show that the last liberality objection doesn't succeed. Since GSE is more principled than the traditional selected effects theory, and it's not too liberal, we should accept it.

CHAPTER 7

Proper Functions Are Proximal Functions

The heart beats. In beating, it moves blood around the body. This brings nutrients to cells and eliminates waste – essential activities for keeping us alive. Which of these activities, precisely, is its function? Are all of these activities its functions, or just one? And if only one, which? Does it matter?

Different people might answer this question differently. One person might say that the function of the heart is just to beat: that is, to expand and contract. The other activities, like moving blood around, or bringing nutrients to cells, are beneficial effects of that function. (I think this is the right view.) Someone else might say that the heart's true function is to circulate blood. The beating is just a mechanism by which it achieves this function, and the other activities, like bringing nutrients to cells and survival, are its beneficial consequences.

The possibility of disagreement stems from a form of function indeterminacy, which I'll call the "hierarchical" form. (Shortly, it will be clear why.) Whenever a trait has a function, we can analyze that function in terms of a sequence of activities, such as beating, blood circulation, bringing nutrients to cells, and survival (see Figure 7.1). It then makes sense to ask: Which member of that series – if any in particular – is its real function? (A trait might have more than one function, such as the male urethra's function of transporting urine and semen. Also, which of the multiple functions is called for at a given time depends on the situation the trait is in – see Garson and Piccinini 2014. None of those complications affect the basic outlook here.)

The problem is compounded by the fact that neither the traditional selected effects theory, nor GSE, tells us which to choose. Consider natural selection between organisms. Here, the function of a trait is the reason it evolved by natural selection, but why did the heart evolve by natural selection? Because it beat, and thereby moved blood around, and thereby brought nutrients to cells, and thereby helped certain of our ancestors, the ones with hearts (or proto-hearts), live longer than those without hearts.

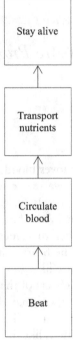

Figure 7.1   The hierarchical form of the function indeterminacy problem.

All of those activities are its selected effects. If functions are selected effects, all of those activities are its functions. There's no principled way of putting one on a pedestal and calling it the heart's "real" function.

GSE isn't alone here. Most theories of function face the same problem. Consider Boorse's biostatistical theory (BST). For BST, the function of a trait is its statistically typical contribution to fitness. It's just as correct to say that the heart's beating contributes to fitness as it is to say that the heart's circulating blood contributes to fitness. In BST, both activities are functions.

The problem is mitigated for the causal role theory (CR) of function. That's because CR has an explicitly perspectival edge. In CR, the function of a trait is its contribution to a capacity that someone deems particularly interesting. Functions depend on the mind. Cummins (1975, 762) writes that functions are always relative to an "analytical context"; Craver (2013, 142) states that functions are ultimately defined by reference to "observer interests and perspectives." Maybe some of this indeterminacy can be whittled down by the observer's interests. If an observer is sufficiently precise about which activity he or she cares about – say, blood circulation versus pushing nutrients around – then he or she can give precise functions to all of the parts of the system. However, I'm not interested here in BST or CR since they don't capture function's explanatory depth. I want to know whether we can solve function indeterminacy from inside GSE.

Why does indeterminacy matter? Who cares which of these activities is the heart's function? Why can't all of them be its functions? There are actually two different versions of this catholic idea, the pluralist and the contextualist. The pluralist says: All of these activities are the heart's functions. Beating is a function, and so is circulating blood, and bringing nutrients to cells is, too. This is Neander's (1995b, 119) official position. The contextualist says: Which of these activities is the heart's function depends on the context of inquiry (Goode and Griffiths 1995). For example, since the neuroanatomist and the behavioral ecologist have different interests, then different ways of describing a trait's function will be appropriate to each. At first glance, these are respectable enough positions. Most of us have better things to do than squander our time on philosophical pseudo-problems. We want to be sure that there is a right answer, and that something depends on getting the answer right.

This chapter unfolds in five sections. First, I'll motivate the problem by addressing why indeterminacy matters for biomedicine and teleosemantics (Section 7.1). Then I'll set out a variant on Neander's solution that says that the function of a trait is its most proximal activity, giving two arguments for it (Section 7.2). Section 7.3 will use the example of the neuromuscular junction to show how the solution works and deal with another objection. Proper functions are proximal functions, but can we spell out this idea in any more detail? Sections 7.4 and 7.5 complete the task. Section 7.4 discusses Neander's view that the function of a trait can be specified very precisely if we situate it in the context of *functional analysis*. In the last section, I argue that we can do even better by situating a trait's function in the context of *mechanistic analysis* (Section 7.5). Crucially, mechanistic analysis isn't supplanting GSE but supplementing it.

7.1 What Function Indeterminacy Is and Why It Matters

There are two reasons that function indeterminacy matters. The first has to do with biomedicine and the second with teleosemantics. First, solving function indeterminacy is useful for giving a precise characterization of when something dysfunctions. If nobody can say exactly what something's function is, then nobody can say exactly when that thing dysfunctions.

Consider how different attributions of function can lead, in theory, to different attributions of dysfunction. Suppose we say that the function of the heart is to circulate blood, not to beat. What if a person is having a brain hemorrhage, and the heart can't circulate enough blood to keep the person alive? In other words, it can beat just fine; the problem is rapid blood loss due to a broken brain artery. Is the heart dysfunctional or not? Technically, because the heart cannot perform its function, we might be inclined to say the heart is dysfunctional.

What if, instead, we say the function of the heart is just to beat – that is, to expand and contract? (I realize the situation is a bit more complex, since the heart must also modulate its rate in order to maintain a constant ratio of carbon dioxide to oxygen; I'll ignore such niceties at present.) Then, in the case I have described, the heart is not dysfunctional. It's doing exactly what it's supposed to do. My point is that different ways of describing the heart's function suggest different conclusions about dysfunction; so, function indeterminacy matters for biomedicine.

I realize the heart is a very simple example. Someone might say: "Look, even if you say that the function of the heart is to circulate blood, and the heart can't do that because of a ruptured brain artery, nobody would say that the heart is dysfunctional. It's obvious that the dysfunction is with the artery. You don't really need to solve function indeterminacy to figure out when the heart dysfunctions. The problem of function indeterminacy and the problem of dysfunction should be kept apart."

I can't accept this line of reasoning. The example of the heart shows that when it comes to assessing dysfunction, we implicitly have a certain solution to function indeterminacy in mind. In other words, it's not that we don't need a solution to function indeterminacy; rather, we already have the solution and just need to spell it out. It's obvious that the heart isn't dysfunctional, because we can see that it performs its function just fine, but that means we're already privileging one of the heart's activities over another as being its function: namely, the lowest-level activity of beating. I'll come back to this point soon.

Unfortunately, not all cases of dysfunction are as clear-cut as the heart example. Consider a case of filial imprinting gone awry. Filial imprinting is an early developmental window in some animals where they develop a strong, lifelong attachment to their own parent, usually the mother. The way it works in goslings is that the gosling imprints on the first large, moving object it sees, within the first two days of its life. In the typical case, the first large, moving object it sees is its mother. Filial imprinting goes awry when something else pops into the gosling's view at the pivotal moment. One of the most famous pictures of the ethologist Konrad Lorenz shows him smoking a pipe followed by a flock of goslings. He discovered that they had imprinted on his boots.

Here is the question: In the case of Lorenz's boots, did anything dysfunction? Is the imprinting mechanism itself dysfunctional? Is the product of that mechanism dysfunctional: namely, some synaptic configuration that makes the gosling follow around a pair of boots? In these cases, unlike the heart case, intuitions get muddy. If the function of the imprinting mechanism is to cause the gosling to form a strong attachment to the first large, moving object it sees, there's no dysfunction. The imprinting mechanism discharged its function admirably when the gosling imprinted on a pair of boots. If the function of the imprinting device is to cause the gosling to follow around its mother, then we have a strong case for dysfunction. A good solution to function indeterminacy should guide our judgment in these puzzling cases. This problem will come up again in Chapter 11 when we think about mental disorders.

Here's a real-life example where poor attributions of dysfunction can actually do grave harm. In the 1970s, the main neurochemical theory of schizophrenia was the dopamine hypothesis (Garson 2017c). In this view, crudely put, schizophrenia stems from too much dopamine in the brain. One strength of this theory is that it accounted well for the antipsychotic properties of the dopamine-blocking medication chlorpromazine. It also tacitly justified that approach to treatment: If schizophrenia is caused by hyperactive dopamine neurons, then the most rational thing to do is to pummel the brain with dopamine antagonists.

In the 1990s, however, the dopamine hypothesis began to unravel. First, despite two decades of ceaseless effort, direct biochemical evidence for the dopamine hypothesis remained elusive. Second, the effectiveness of so-called atypical antipsychotics, which target a much wider profile of neurotransmitters, suggested that dopamine might be one little piece of a much bigger puzzle. A handful of researchers went so far as to suggest that dopamine abnormalities in schizophrenia might have the *function* of compensating for defects in other neurotransmitter systems (Grace 2000, 332). If that's correct, then dopamine changes associated with schizophrenia wouldn't be dysfunctions at all. They would have functional and adaptive significance, just like swelling or coughing. Bombarding dopamine receptors with chlorpromazine is not only potentially bad for some people; it also makes them vulnerable to side effects like painful and debilitating movement disorders.

There are three morals we should draw from this story. First, it can be difficult to pinpoint the locus of dysfunction in a complex system. Second, it matters how we do so. Third, solving function indeterminacy is an important part of that task.

As I alluded to above, there is a second reason function indeterminacy matters, outside of biomedicine. This has to do with teleosemantics, the view that mental representation is based on biological function. Roughly, teleosemantics says that the content of a mental representation, what it's about, depends on the function of the mechanism that makes or uses it. As a consequence, any lingering indeterminacy about functions might carry over into lingering indeterminacy about contents, depending on which version of teleosemantics one holds. I will return to the topic in Chapter 12.

7.2   Distal and Proximal Functions

Everything I've said so far goes to show why indeterminacy is a real problem. Now for the solution. The question is this: Do we have any principled (that is, nonarbitrary) and context-independent (that is, not dependent on this or that scientific discipline) way of picking just one activity, in the sequence of activities, as the trait's real function? For example, do we have any principled basis for saying that the heart's real function is to beat, rather than to circulate blood or to bring nutrients to cells?

We do; we have good reasons to say that the function of a trait is the activity that's most *proximal* to that trait. In other words, the function of the trait is just the very first activity in the long sequence. The function of the heart is to beat, not to circulate blood. The circulation of the blood, to be strict, is a beneficial result of its performing its function, not the function itself. To put the point in a simple and memorable slogan, *proper functions are proximal functions.* (Hence the chapter's title.)

There are three arguments for this point of view. I'll call them the arguments from *intuition, intervention,* and *informativeness.* The first relies on a certain intuition about when things dysfunction. Go back to the ruptured artery scenario. Here, the heart beats just fine, but, because of a ruptured blood vessel in the brain, it cannot circulate enough blood to keep the person alive. Is it dysfunctional or not? Intuitively – if you and I share the same intuitions – it isn't dysfunctional. The heart is doing exactly what it's supposed to do. If anything's dysfunctional, it's the blood vessel.

Suppose, instead, the heart stops beating. It simply cannot contract and expand. Is it dysfunctional then? Intuitively, it is. It's worth taking a long pause to contemplate this peculiar asymmetry in our intuitions about when things dysfunction. If the heart is beating, but it isn't circulating blood, then it's not dysfunctional, but if the heart stops beating, it is. What's going on?

Contrary to first appearances, we do seem to have a principled and context-independent way of picking out a specific activity in the sequence of activities. As a rule, and with a qualification to be addressed at the end of the chapter, we should say that the function of a trait is that activity the inability to perform which implies dysfunction of that trait. Papineau (1997, 4) puts the point in a very similar way. The inability to circulate blood might or might not involve a heart dysfunction. It might be the heart; it might be the artery. Since the inability to beat does imply heart dysfunction, the function of the heart is to beat, not to circulate blood.

Note that, while this solution is inspired by Neander's work, it is not Neander's solution. Neander (1995b, 118–119) accepts a pluralist stance on function indeterminacy: "it is also consistent with the definition of 'function' being employed here that it is the function of the trait to *help* achieve the higher functions." She just thinks that when function attributions lead to conflicting results, we should stick with the more proximal activity (ibid.). I am going further by rejecting pluralism. The function of the heart is just to beat, and not to circulate blood. There are no distal functions; the corresponding attributions ("the function of the heart is to circulate blood") are plain wrong. Specifically, distal function attributions commit the fallacy of division. They identify a function of a whole (in this case, the circulatory system) and attribute it to a part (the heart).

This intuition about functions, where proper functions are proximal functions, finds expression in Dretske's famous magnetosome case. Dretske (1986, 26) describes a kind of anaerobic bacterium that has an inner magnet that aligns it with geomagnetic north. This magnet leads it away from the oxygen-rich surface water that would otherwise kill it. Should we say that the function of the magnet is to align the bacterium with the prevailing magnetic field (proximal) or to keep it away from oxygen-rich water (distal)? Dretske thinks we should stick to the proximal level of description. He says that if we hold a bar magnet above the bacterium, it'll swim toward the magnet, and hence toward the oxygen-rich water, and die. In this case, he thinks, nothing is malfunctioning. The magnet discharged its function perfectly well when it caused the bacterium to move toward the bar magnet. That's the intuition that my first argument hangs on.

There's a problem with this argument. The issue is that people's intuitions differ about when things dysfunction. Maybe some people are comfortable saying that in the ruptured blood vessel case the heart is dysfunctional. Fortunately, I can go beyond intuition to show why we shouldn't think that way. This is the second argument, and it is based on Buller (1997). The argument has a more pragmatic flavor. From a practical perspective, when we say that a trait dysfunctions, we're suggesting that it's an appropriate target of medical intervention. It's the sort of thing one might wish to fix, repair, or replace to get the system up and running again. However, if the heart cannot circulate blood because of a ruptured vessel, we don't want to fix the heart. We want to fix the artery!

Neander makes a similar argument when she appeals to the informativeness of dysfunction talk. I consider this a third argument. She thinks we should only say that a trait dysfunctions when it can't perform its most proximal function, since that makes dysfunction talk highly *informative* (1995b, 120). Suppose I'm told that someone's circulatory system is dysfunctional. Suppose I'm told, later on, that the person's heart is dysfunctional. Then (on Neander's preferred way of speaking) I've learned something substantially new. I've learned that the heart is what's causing the circulatory system to fail to perform its function. If we use the term "dysfunction" loosely – and we say that the heart is dysfunctional only because it can't perform one of its distal functions (say, contributing to circulation) – then when I'm told the heart is dysfunctional, I don't learn anything new. All I learn is that some high-level system of which the heart is a component isn't performing its function, but I'm still in the dark about what's causing the problem.

In sum, we ought to use the term "dysfunction" narrowly, and say that a trait dysfunctions only when it can't perform its most proximal function. If we accept that, then we do have a principled and context-independent solution to function indeterminacy. That is, we have a nonarbitrary basis for picking out one specific effect in the whole series of effects associated with a trait's function and identifying it has the proper function of that trait: namely, its most proximal effect.

7.3 An Objection to Proximal Functions

Here's one more example to illustrate the proximal solution to indeterminacy. It also forms the basis for a new objection. In the mammalian neuromuscular junction, motor neurons from the brain synapse onto muscle fibers rather than other neurons. Now, consider a motor neuron. What's the function of that neuron? Think about the sequence of activities by which the neuron historically contributed to the creature's fitness. The neuron released acetylcholine (ACh). By doing so, it activated receptors on the muscle fiber. By doing so, it caused a muscle contraction. By doing that, it helped us move our bodies. Which of these activities is its function? If proper functions are proximal functions, then the function of the motor neuron is just to release ACh. All the rest are, speaking strictly, beneficial consequences of the performance of that function.

This solution matters when we think about various diseases of the neuromuscular junction. One disease is Lambert-Eaton syndrome. It is an autoimmune disorder that stops the motor neuron from releasing ACh. If a neuron cannot release ACh due to Lambert-Eaton syndrome, it is dysfunctional. That is because it cannot perform its most proximal function for constitutional reasons (the next chapter will go deeper into what dysfunctions are).

Consider another disease of the neuromuscular junction, myasthenia gravis. Myasthenia gravis doesn't impair the motor neuron. It stops the muscle fiber from binding to ACh. That is, the motor neuron releases neurotransmitter just fine, but the muscle cannot bind it. It is just like the heart and the ruptured vessel. Strictly speaking, in the case of myasthenia gravis, the motor neuron is not dysfunctional. It is doing exactly what it is supposed to do. Instead, we should say the muscle fiber is dysfunctional. A good solution to function indeterminacy should help us make wise judgments in complex biomedical contexts.

One might object to this solution to function indeterminacy by pointing to how biologists routinely talk. I said the function of the motor neuron is to release ACh, but I imagine that if you ask physiologists what the function of the motor neuron is, they'll say its function is to contract muscles, which is a more distal effect. Physiologists are happy to cite distal effects as functions. To round out the objection, you might even think that this more distal way of talking is actually quite informative and natural on some occasions. If my ten-year-old son asks me what motor neurons are for, I would probably tell him that their function is to contract muscles. In other words, I would identify its function with a distal effect. Something has gone wrong with my solution to indeterminacy if it runs against how biologists and even ordinary people like to talk.

Nonetheless, I won't budge from my position. To say "the function of the motor neuron is to cause a muscle contraction," is, strictly speaking, incorrect. It is loose talk. What should I say to my son if he asks me why we have motor neurons? If I wanted to be extremely precise, I would describe muscle contraction as one benefit, among others, of the motor neuron's function. Another way I could answer him is by saying that the function of the motor neuron is to do something that contributes to muscle contraction. I'm not saying its function is to cause a muscle contraction; rather, its function is to perform some unspecified activity that contributes to contraction. Third, I might just tell him that its function is to cause a muscle contraction, knowing that it is, in the end, a harmless little lie.

One might try to criticize this position on the grounds that some traits, like the male urethra, have multiple functions. I seem to be insisting that a trait only has one function. However, I'm not saying that. Nothing I've said prevents a trait from having more than one function. For each of a trait's functions (such as transporting semen and transporting urine) there's a separate chain of events. For the first, by transporting semen, the urethra helps fertilize females, which enhances reproductive success. My view is that only the first of these (transporting semen) is a function; the rest are beneficial effects. For the second, by transporting urine, the urethra disposes of bodily wastes, which prevents toxicity-related diseases. Again, only the first member of the series (transporting urine) is a function. I grant that a trait can have more than one function; it's just that each of those multiple functions should be described in a proximal way.

7.4   Functions and Functional Analysis

A proper function is a proximal function. It is the activity that is closest to the trait in the sequence of activities. Can we say anything more precise about this proximal function? Can we define it more rigorously?

Neander (1995b, 118) developed this idea by saying that the most specific activity of a trait is the one that it can perform "more or less on its own," rather than "in collaboration with other components." Yet this isn't quite right, since the heart doesn't do anything on its own, as Neander (2013, 35) acknowledges. It needs the help of other organs and processes, like electrical impulses from the vagus nerve, and an adequate supply of oxygen and blood.

Neander amplified her comments by putting them in the context of *functional analysis*. Functional analysis works in the following way (see Cummins 1983). First, pick out a high-level capacity of a system, like a creature's ability to survive. Then explain that capacity by analyzing the system into a number of sub-capacities, such as breathing, digestion, and heat retention. Then show how these sub-capacities add up to yield the higher-level capacity. This process is iterative. We can show how a given sub-capacity performs its function by dividing it up into a number of sub-capacities, and so on. Suppose we want to explain survival, and analyze it into sub-capacities like breathing and digestion. We can then home in on a specific sub-capacity, like digestion, and analyze that into its sub-capacities, such as mashing food, absorbing nutrients, and eliminating wastes.

Functional analysis gives us a neat and accurate way of thinking about function indeterminacy. Neander (1995b, 118) observed that as we move up through the sequence of activities associated with a trait's function, each activity is performed by a more and more inclusive system (see Figure 7.2). Consider the lowest level activity in the hierarchy of functions – beating. Which system performs this activity of beating? The heart performs this activity (though not all alone). Consider the next activity in the hierarchy, circulating blood. Which system performs this activity? The circulatory system does. The heart is just one part of the circulatory system, but there are other parts, too, like the veins and arteries, that all work together to circulate blood. Consider the next activity in the hierarchy – bringing nutrients to cells. This activity is performed not only by the circulatory system but also by the lungs, which bring oxygen into the blood stream and release waste. Finally, the entire group of organ systems work together to "perform the activity" of keeping the creature alive.

This is why it's appropriate to label the specific form of indeterminacy we're considering as the *hierarchical* form of indeterminacy. Each activity in the sequence has its own level in a hierarchy of nested systems. It also helps us understand why, when we identify a trait's function with a distal effect, we're committing a fallacy of division. The function of circulating blood does not belong to the heart but to a bigger system that includes the heart among its parts.

Once we have this hierarchical picture in mind, we have a very convenient way of identifying a trait's function. Its function is just the activity that emerges at the lowest-level of a functional analysis, when the relevant hierarchy of activities (e.g., beating, circulating blood, bringing nutrients to cells) has been identified by GSE. The function of the heart is to beat

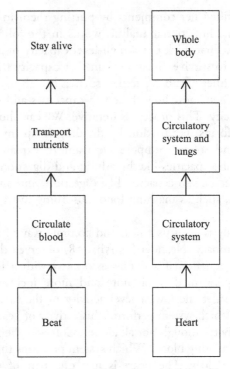

Figure 7.2   Functional and systemic hierarchies.

because that's its specific activity (or capacity) in a functional analysis of how the trait contributed to its differential reproduction or differential retention in a population.

You might wonder how the solution given here, where the function of a trait is its specific activity in a functional analysis of this sort, goes beyond the sort of analysis that the causal role (CR) theory of function gives on its own. If you recall, CR says that the function of a trait is just its contribution to some interesting system-level capacity, and that we should use functional analysis to identify that contribution.

This solution goes far beyond CR. That is because we're using GSE to pick out the relevant hierarchy of activities. GSE gives us a rationale for picking one particular sequence of activities (beating, circulating blood, bringing nutrients to cells) as being relevant to the heart's function. We then use functional analysis to move from indeterminately specified functions to determinately specified functions. CR, all alone, makes functions depend on our interests and goals, giving no uniform standard for picking out one hierarchy over another. Relative to an interest in lie detectors, the function of the heart is (supposedly) to cause a spike in blood pressure. Relative to an interest in medical diagnosis, the function of the heart is to make beating sounds one can listen to through a stethoscope. By firmly subordinating functional analysis to GSE, we solve function indeterminacy without relying on perspectives and conventions.

7.5 Functions and Mechanistic Explanation

I want to close the chapter by saying something about functions and mechanistic explanation, although I'll have more to say on the topic of mechanisms in Chapter 10. Neander's solution to function indeterminacy, which deploys functional analysis to identify a trait's "most specific" function, can be fruitfully recast in the more recent framework of mechanistic explanation, as philosophers of science have developed that framework over the last two decades. This framework helps to make Neander's solution to indeterminacy more precise. This framework, in its modern form, emerged during the 1990s and became prominent over the last two decades (e.g., Bechtel and Richardson 1993; Glennan 1996; 2017; Machamer, Darden, and Craver 2000; Darden 2006; Craver and Darden 2013).

One of the themes of the new mechanism tradition is that scientific progress consists largely of the discovery of mechanisms for things. One important sort of mechanistic explanation, but not the only one, is reductionistic: We show how a system produces a phenomenon by identifying the parts of the system and the activities of those parts. We then show how the phenomenon in question arises given the way those parts and activities are organized (e.g., Bechtel and Abrahamsen 2005, 426). Sarkar (1998, 45) describes a similar procedure in terms of giving a "strong reduction." There are four ideas that are crucial for reductionistic mechanistic explanations: phenomenon, components, organization, and hierarchy. I'll take each in turn.

A mechanism is always a mechanism for a phenomenon. The phenomenon is the thing we're trying to explain. The question "How many mechanisms are in the human body?" doesn't make sense, since it doesn't say which phenomenon must be explained. Once we have a preliminary "fix" on the phenomenon, we then examine the system that produces the phenomenon, and analyze it into its components.

By "components," I mean both the physical parts of the system, and their characteristic activities. This emphasis on components is one feature that distinguishes mechanistic analysis from functional analysis, as Craver (2001) and Piccinini and Craver (2011) emphasized. According to Cummins (1983, 29), when we analyze a capacity into a number of sub-capacities, these sub-capacities may or may not correspond to real physical parts of the system (until we reach the bottom level of the functional analysis). In a mechanistic analysis, the system's components are always real physical parts, at every level of analysis.

The third important feature of mechanistic explanation is organization. We explain a phenomenon by showing how the various components are organized. Specifically, we reveal their spatial and temporal relationships and their physical connections. This is a vast improvement over Cummins' rather vague insistence that in functional analysis we show how the sub-capacities are "organized in a way that could be specified in a program or flow chart" (1975, 579, fn 16).

The final component is hierarchy. Mechanisms are, except in the simplest of cases, hierarchically organized. Mechanisms have mechanisms inside them, until we reach the ground floor of fundamental physics. For example, suppose I describe the mechanism of the action potential. For my purposes, I divide up the neuron into four parts: dendrites, soma, axon, axon terminal. I describe the action potential as consisting in the following process: the dendrites transmit electrical impulses to the soma; the soma adds up those charges and, if they pass a certain threshold, produces an electrical impulse; the axon carries that impulse to the terminal; the terminal releases neurotransmitter. Suppose we want to know about how the axon terminal releases neurotransmitter. Then I treat the axon terminal itself as a mechanism in its own right, the activity of which is to release neurotransmitter, and I carry out a similar style of analysis to show how it does so. The fact that mechanisms are hierarchically embedded gives rise to a project called "multi-level mechanistic modeling" (Darden 2006).

One benefit of recasting Neander's solution in this framework is that it allows us to define the notion of function in a highly precise way. In short, and with some qualifications to follow, I recommend the following char-acterization: the function of an item is its (historical) contribution to the activity of the mechanism in which it is immediately contained, when this hierarchy of activities has been identified by GSE. For example, the function of the heart is to beat, because that is its contribution to the activity of the mechanism in which it is immediately contained, namely, circulating the blood, when the relevant hierarchy of activities has been identified by GSE. As noted above, mechanistic considerations aren't supplanting GSE but supplementing it. GSE yields indeterminate functions; mechanistic considerations yield determinate ones when they're added on top.

Another benefit of recasting her solution this way is that it gives us a rigorous way of thinking about dysfunction. The problem, which I'll discuss in the next chapter, is this: Even when a trait can't perform its most specific function, that doesn't mean it's dysfunctional. It might just not have the resources, or "fuel," it needs to perform its function. Consider the heart. Earlier, I noted that even though the function of the heart is to beat, it cannot beat all by itself. It needs certain resources. It needs a constant supply of blood. It also needs impulses from the vagus nerve. The mechanistic framework gives us a way to model the interaction of components that the heart needs to be able to perform its function.

CHAPTER 8

When Functions Go Wrong

What are dysfunctions? Sometimes we talk about dysfunctions for artifacts, such as "wardrobe malfunctions" and "manufacturer defects." Here,
the idea of dysfunction is easy to grasp: Something dysfunctions, roughly,
when it doesn't work as it's meant to. (I assume it's a bit more complex,
but intention strikes me as a great starting point for figuring out how
artifact functions work; see Krohs and Kroes 2009 for a recent anthology.)
Is the idea of a biological dysfunction anything more than a metaphorical
extension of the idea of an artifact dysfunction?

Some go so far as to say that there are no dysfunctions in nature,
outside of our goals, projects, and plans. The psychologist Peter Sedgwick
(1981, 121) put it well when he announced that "outside the significances
that man voluntarily attaches to certain conditions, *there are no illnesses or
diseases in nature.*" When we label something a "disease," he thought, we
express a *disvalue* – we judge the condition to be undesirable or bad. In
philosophy of medicine, this outlook is called "normativism." Maybe we
should think about dysfunction in the exact same way, as an all-too-
human projection of disvalue upon nature's impassive face. To butcher
Hamlet, perhaps there are no dysfunctions in nature, but thinking
makes it so.

At an extreme, one might think that the very idea of dysfunction – and
its gang of associates like defective, disabled, and deformed – is politically
toxic. It is true that sometimes, when we label things "dysfunctional," we do
nothing more than to express our opinion that it's bad and it ought to be
changed. Consider "dysfunctional families" or "dysfunctional relationships."
Today, many argue that some of our alleged disabilities, like being deaf or
hard of hearing, are not disabilities at all; they are normal variations on how
bodies are. As Barnes (2016, 6) puts it, "to be disabled is to have a minority
body, but not to have a broken or defective body." To call those conditions
dysfunctional might drag in a host of oppressive value judgments, and it

might lead to the further marginalization of already marginalized groups (Amundson 2000).

I agree that we ought to be vigilant about calling things "dysfunctional." In Chapter 11, I'll argue that psychiatrists often throw around the "dysfunction" label without considering the prospect that the conditions so labeled, whether autism, anxiety disorders, or the delusions of schizophrenia, aren't actually so. Coming to grips with what functions and dysfunctions are is a powerful tool for criticizing the oppressive use of those labels, while respecting their appropriate domain.

I want to insist, however, that they do have an appropriate domain. Even Amundson (2000, 34), who's otherwise very critical of the "dysfunction" label, says his critique is mainly focused on fairly permanent and stable conditions ("disabilities") rather than episodic or life-threatening ones like cancer or measles ("diseases"). That suggests, to my mind, that he's open to the idea that cancer, measles, or strokes involve real dysfunctions. My advice? Apply it where it deserves to be applied, and be ruthless against its facile overextension.

This chapter uses GSE to say what dysfunctions are, and irons out some complexities that arise along the way. I'll begin the chapter by distinguishing two ways a trait can fail to perform its function: one due to the trait's constitution, the other due to its circumstances. Strictly speaking, a trait dysfunctions only when the failure is due to its constitution, not its circumstances (Section 8.1). I then argue that in order to understand what dysfunctions are, we have to understand the corresponding idea of a trait's normal environment (Section 8.2). Something dysfunctions just when it can't perform its most proximal function in its normal environment. I conclude by making another pitch for GSE. A strength of GSE is that it explains function's normativity in an elegant way; I contrast this with Boorse's biostatistical theory, which can explain dysfunction, if at all, in an extremely tortured way (Section 8.3) – and what comes out of it still runs against ordinary biomedical usage.

8.1   Constitution and Circumstance

The key to GSE's (and the traditional selected effects theory's) ability to account for dysfunction is that it invokes history. History lets us sharply divide something's having a function from something's being able to perform it. Having a function, for GSE, depends only on history. The ability to perform it depends only on current-day constitution. It's easy to see how they come apart. All etiological theories of function have this advantage, not just the selected effects theory. For example, the "persistence-plus" theories I outlined in Chapter 3 can account for the possibility of dysfunction. The main problem is that they're too liberal.

Consider, in contrast, Cummins's (1975) starter version of the causal role (CR) theory. He is quite explicit, there, that for a trait *token* to have a function, it must be able to perform that function: "function-ascribing statements imply disposition statements; to attribute a function to something is, in part, to attribute a disposition to it" (758). Cummins's view implies that if a trait token cannot do something – that is, if it has lost the disposition entirely – then that activity isn't its function, nor can there be any dysfunction. Some of CR's proponents have tried to remedy this problem, such as Hardcastle (1999) and Craver (2001), although they still come up short (see Garson 2016, Chapter 5 for an overview and critique).

Boorse's biostatistical theory (BST) presents a somewhat more promising avenue for grasping dysfunction, but it still has problems – three, to be exact. These are problems that GSE entirely skirts, but I will not make my case for that right now. Instead, I'll set out what GSE says dysfunctions are. Then, at the end of the chapter, I'll work my way back around to Boorse's BST. We can compare and contrast the two views when both have been carefully examined.

In the last chapter, I showed how solving function indeterminacy is *necessary* for deciding when something dysfunctions. In other words, if we don't know exactly what something's function is, we don't know exactly when it dysfunctions either. It's tempting to think that solving indeterminacy is *sufficient* for deciding when something dysfunctions, too. In other words, you might think that by solving indeterminacy, we also figured out, without further ado, what dysfunctions are. Something dysfunctions just when it cannot perform its most proximal function. (Remember, in my view, functions are just proximal functions, but sometimes I'll still specify that I'm talking about proximal functions because it's crucial to keep that in mind but easy to forget.) The heart dysfunctions only when it cannot beat. The motor neuron dysfunctions only when it cannot release ACh.

Sadly, being unable to perform something's most proximal function isn't quite enough for dysfunction. There needs to be something more. The point is obvious in the case of artifacts. Consider an unplugged toaster. It cannot perform its function, no matter how proximally that function is described (say, to heat some coils when a lever is pressed), but it isn't dysfunctional. Someone just forgot to plug it in. The mere inability to perform one's most proximal function cannot be enough.

Although artifact functions aren't biological functions, the same lesson applies to biology. A function of the eye is to see: that is, to process information about light. If I'm blindfolded, I cannot see. My eyes cannot perform their function, no matter how proximally that function is described – but they aren't dysfunctional. What did we leave out?

Here's what we left out. When something cannot perform its function, there is one of two causes: the constitution of the thing or its circumstances. Put crudely, either the thing itself is "broken," or the environment is "uncooperative." The item is only dysfunctional in the first of these cases. In the toaster and blindfold cases, the item cannot perform its function, but the fault is entirely with the circumstances, not the constitution (Dretske 1986, 29). The toaster is only dysfunctional when it cannot perform its function and it's in its normal operating circumstances – namely, hooked up to a power source. The eye is only dysfunctional when it cannot perform its function and it's in *its* normal operating circumstances – it's hooked up to a light source. Millikan (2013, 40) put the point succinctly in a recent paper: "When a device does not perform some proper function only because the necessary background conditions are absent, we do not consider that to be a malfunction. . .Malfunction results only from abnormalities in the constitution of the device itself."

With this rough distinction between constitution and circumstance in mind, it's easy to say what dysfunctions are. Something dysfunctions (with respect to some function) if and only if it cannot perform its most proximal function, for constitutional reasons. Still, all of this needs to be spelled out in detail.

First, a short digression on the word "cannot." One might say, "The idea that dysfunction has to do with constitution is already implied in the word 'cannot'." When we say something cannot perform its most specific function, we're saying it has entirely lost the disposition, or capacity, to perform the function, not just that it's in an unsuitable environment. For example, in the unplugged toaster example, I wouldn't say that the toaster cannot toast bread. I'd say that it can toast bread, since all of its inner parts are arranged just as they're supposed to be. Therefore, the second part of your definition, where you say it has to be for constitutional reasons, is redundant."

I don't want to linger long on the nuances of the word "cannot." I admit that, to my ears, it seems obvious that an unplugged toaster cannot toast bread. Just try it! I do understand perfectly well, however, if someone wishes to use the words "can" and "cannot" in a richer sense, involving the having or lacking of certain dispositions. If that is what we agree to mean, then I agree that something dysfunctions only when it cannot perform its most proximal function. We still, however, have the burden of unpacking exactly what this richer sense of "cannot" amounts to. In the following, I'll continue to use the word "cannot" in the unsophisticated way (that is, in the sense in which an unplugged toaster cannot toast bread), and I'll see what has to be added to get a dysfunction.

When something cannot perform its function, whether a broken heart or an unplugged toaster, the cause is either in the constitution of the thing, or its circumstances (or both, but I'll ignore the hybrid case, since it won't matter anyway). We should clarify what it means for the environment to be "uncooperative."

On the surface of things, it would seem that there are two ways for the environment to fail to cooperate. First, we might (erroneously) specify the function of a trait very distally. In other words, we might commit the fallacy of division and attribute a function of a whole (circulating blood, which is a function of the circulatory system) to a part (the heart). Now, suppose the heart is beating, but it cannot circulate blood owing to a ruptured brain artery. Then it cannot perform its "function," but it isn't dysfunctional. This sort of case doesn't really exist. It only seems to be real because we failed to identify the item's function correctly. It's a consequence of a misleading way of speaking and thinking about functions.

Even when we restrict our attention to most proximal functions, there is a genuine way the environment can be uncooperative. This is when the trait lacks the *preconditions*, resources, or fuel it needs to do its job. Let's go back to the unplugged toaster. Even when we restrict our attention to its most specific function (warming coils in response to a lever press), it still doesn't have a resource it needs to perform it – namely, an electrical current. The lesson applies to the blindfold case, too. Even with a proximal way of describing the eye's function (to see, to process information about light), if I'm blindfolded, my eyes cannot do that. A precondition is the availability of light.

How should we flesh out this idea that in a dysfunction the inability of a trait to perform its function has to do with constitution, not circumstance? The best way is to put the burden on the idea of the *normal environment* for a trait's function. The normal environment for a trait's function is that environment in which it historically performed that function, and in which those performances boosted its relative fitness. The normal environment for the eyes' performing the function of seeing is one with a lot of light around, and in which they are attached to a brain and a body, and in which there are things, both moving and still, for it to see. That was when having eyes capable of seeing made a real difference to our ancestor's survival prospects. The perceptive reader will note that my idea of a normal environment echoes Millikan's (1984, 33) notion of the *normal condition* for a trait's functioning. "Normal," here, doesn't depend on what's statistically average for a trait. A trait might have performed its function rarely; its normal environment would still be the environment in which it did perform that activity and in which those performances boosted its relative fitness.

I will say more about the idea of a normal environment, but the important point is this. Once we have the idea of a normal environment in place, then if we want to know whether or not a trait token is dysfunctional at any moment, we ask ourselves the following question: If that token, given its exact constitution at that moment, were placed in its normal environment, would it perform its function? If so, it is not dysfunctional. The blame for the failure falls on the environment. If not – that is, if the trait were placed in its normal environment and it still couldn't perform its function – then the trait is dysfunctional. The blame for the failure falls on the constitution.

Let's do a dry run to see how this works for the eyes before we get into tougher cases. The normal environment for the eyes' seeing is one equipped with light, one where the eyes are attached to a body and brain, and one where there are things around to be seen. Suppose I cannot see. Are my eyes dysfunctional? What would happen if we took those eyes, with their exact constitution, and put them in their normal environment? Would they be able to see then? If so, there is no dysfunction. That's why, if I'm merely blindfolded, my eyes aren't dysfunctional. If you put my eyes back in their normal environment, they would see just fine. In contrast, if I cannot see because of late stage retinitis pigmentosa, my eyes are dysfunctional. If you put my eyes back in their normal environment – but without changing anything about their constitution – they still would not see.

8.2 Normal Environment and Selective Environment

There's still a tricky question to resolve. Which features of the environment, exactly, are in a trait's normal environment? In other words, if I want to know whether or not a trait is dysfunctional, I have to be able to describe, in some detail, what this normal environment is like. What has to be included in that description? (Put differently: if we were to recreate the trait's normal environment, which features would we have to add?)

Consider language use. Suppose that language evolved in humans about 100,000 years ago. What is its normal environment? At that time, humans lived in small, hunter-gatherer bands. Large carnivores like lions, hyenas, and leopards preyed on our ancestors. The atmospheric carbon dioxide concentration was lower than it is today, somewhere between 200 and 300 parts per million. When we describe the normal environment for language use, do we have to include all those things? Some of them, like CO_2 concentration, scarcely seem relevant.

Brandon (1990, chapter 2) gives a useful discussion of the concept of *environment* in biology, and his discussion can help us along. The big question motivating his discussion is this: Suppose a species is spread out over a large, varying region of space and time – say, a bunch of genetically identical sunflower seeds are dispersed over a vast field. Suppose different parts of the region are made of different soil types and are exposed to varying amounts of rainfall. How many different "environments" are there? How do we count the environments in this region?

With this end in mind, he distinguishes between three different senses of "environment:" a trait's *external environment, ecological environment*, and *selective environment*. The external environment for a trait consists of all the environmental factors that affect it. This notion of environment is rather indiscriminate and includes more or less any physical or biotic factors outside of the organism that may or may not affect its survival prospects. Normal variations in atmospheric CO_2, for example, would be part of its external environment, as would soil types and rainfall.

A trait's ecological environment includes only those environmental factors that affect its absolute fitness (its expected number of viable offspring). Suppose that there are two soil types in this region, A and B. Suppose that seeds grown in soil A are fitter than seeds grown in B (again, we'll assume the seeds are genetically identical). Then, a patch of land made of soil type A is a different ecological "environment" (for the seeds) than a patch of land made of soil type B. Suppose, in contrast, that the region can also be divided into two large areas based on rainfall, r1 and r2, where r1 gets a bit more rain than r2. Suppose, moreover, that the difference in rainfall doesn't affect the seeds' fitness. Then, differences in rainfall aren't relevant to the trait's ecological environment. When you're dividing up a region into the seeds' ecological environments, you don't have to mention rain.

The notion of a trait's selective environment is somewhat different from a trait's ecological environment. It's made up of those features of the environment that affect the trait's relative fitness, not its absolute fitness.

To describe the trait's selective environment, you have to consider not only the fitness of that trait but the fitness of any alternative traits. Then, any environmental changes that affect the trait's relative fitness are part of its selective environment. Suppose that there are two different sorts of sunflower seeds, and suppose there is an insect that prefers to feed on the first. Then, the insect is part of the seeds' selective environment. Further, suppose the differences in soil type do not give one kind of seed a fitness boost over the other. Then soil types are not part of the selective environment.

This last notion, that of a selective environment, is the most natural candidate for deciding what to include and exclude from a trait's normal environment. If the function of a trait is just whatever it was selected for, then its normal environment should be made up of precisely those environmental factors that caused it to be selected for. If we're considering the normal environment for language use, we can leave out variation in atmospheric CO_2. That doesn't mean that atmospheric CO_2 isn't relevant to fitness. It means that changes in atmospheric CO_2 don't differentially affect language-users over nonusers. Similarly, suppose that language users didn't avoid predators any better than people who didn't use language. Then the presence or absence of predators wouldn't be part of language's selective environment, since that didn't affect the relative fitness of being a language user. Suppose, however, that the presence or absence of other language users was relevant to the relative fitness of language users (that is, its fitness was frequency-dependent). Then, the presence of other language users would be part of the trait's selective environment.

There are other sorts of questions that arise here. Suppose that, back in the Pleistocene era, hyenas preferred to prey on language users, say, because language users drew attention to themselves with all that jabbering. Then, the selective environment for language use is an environment filled with predatory hyenas. If that's right, most human beings are not, and never will be, in their normal environment for language. One might think that's a problem for how I characterize a trait's normal environment, but I don't see why it would be. If we are never in our normal environments as far as language use goes, that is just fine with me. The crucial thing is that we need to be able to distinguish between function and dysfunction with respect to language use, and the notion of a normal environment is a good conceptual tool for performing that task.

One might think that all this talk of normal environments runs into severe epistemological problems. How does anyone know what the normal environment for language use is? That's a good question, and maybe sometimes it's impossible to know for sure. I take some consolation, however, in three facts. First, my theory says that the function of a trait depends on its recent history, and not its deep history. Recent history will generally be more tractable than deep history. Second, my theory says that selection includes *ontogenetic* selection processes, like learning and synapse selection. Those are generally more accessible to us than evolutionary selection pressures (more on this immediately below). Third, as I noted in Chapter 3, my theory is in good company. Even some of the allegedly nonhistorical theories of function, such as Boorse's biostatistical theory (BST) and Bigelow and Pargetter's propensity theory, make tacit reference to history, and so inherit the same epistemological problems that my theory has. It would be a mistake to think that we should prefer, say, BST over GSE on the grounds that BST makes it easier to assess whether or not something is dysfunctional.

The normal environment for a trait and a function differs depending on what kind of selection process is going on. In the case of ordinary natural selection between individuals, say, the zebra's stripes, normal is the environment where stripes were selected for – in Central Africa over two million years ago. In the case of operant conditioning leading to lever-pressing in a rat, normal is the environment in which that very rat learned to press the lever (a cage with a food source, and so on). In the case of antibody selection, normal is the environment in which the antibody was cloned over others, perhaps during infancy. The fact that the time-scale of normal grows and shrinks, Alice-in-Wonderland style, can make things confusing for determining dysfunction.

Here's a simple example of how our assessment of a trait's normal environment can affect how we identify dysfunctions. (I'll come back to this in Chapter 11, since it causes problems for one popular way of thinking about mental illness.) Suppose a young boy grows up in a conflict zone: For example, an eight-year-old boy is abducted and forced to fight with rebels. Suppose there he acquires dispositions, like being disposed to aggressive outbursts, that serve him well; they bring power and respect, and they are reinforced for that very reason. GSE says that a function of the disposition to aggressive outbursts is to gain power and respect.

Now, suppose he escapes as a teenager. Some of the problems he might face trying to readjust to society have to do with society – will his community still accept him? Others might have to do with the fact that the dispositions that served him well enough with the rebels don't serve him well back home. Now, is there any dysfunction? Are these behaviors or dispositions dysfunctional? (Of course, if all you mean by "dysfunctional" is "harmful," "useless," "counterproductive," and so on, they are definitely dysfunctional, but that is not the sort of function and dysfunction I am talking about.) There, the right answer is "no." To arrive at the answer, one must not ask, do the behaviors serve their function in the new environment? Rather, we ask: Would they serve their function in their normal environment – that is, the environment in which he acquired them? And there, the answer is yes. This is why judgments of dysfunction can be confusing when we don't keep track of the normal environment.

To summarize dysfunction once more, a trait is dysfunctional with respect to some function just in case it cannot perform its most proximal function for constitutional reasons. To spell out the latter, we can say that the trait is dysfunctional if it cannot perform its most proximal function, and, if it were in its normal environment (that is, its selective environment), it still would not perform its function.

This is as good a time as any to address one last problem for dysfunction, which Schwartz (2007) calls the "line-drawing problem." Sometimes, a trait is dysfunctional, not because it cannot perform its function at all but because it performs its function at the wrong rate. The heart dysfunctions not only when it stops beating but when it beats too slowly or quickly. Function grades into dysfunction. Where is the dividing line between them? Schwartz thinks this is "a serious lacuna for any definition of function or dysfunction" (p. 383).

To meet this challenge, Schwartz devises a sophisticated metric for deciding where to draw the line between function and dysfunction. It partly depends on the frequency of the condition in a reference class, and partly on how bad the consequences are. Even after installing this conceptual infrastructure, however, he admits there's still room for "ambiguity" and "arbitrariness" (ibid.) – just not quite as much as before.

I don't see why the line-drawing problem is a problem. It strikes me as an example of the ordinary sort of vagueness that most philosophical concepts have around the edges (Garson and Piccinini 2014, 10). When is a person bald? It is impossible to pin down precisely, and I wouldn't demand of a philosophical account of baldness that it give us a precise demarcation. Philosophers call these "sorites" cases and they seem to affect the whole process of developing concepts and definitions. As Boorse (2002, 71) put it in describing a similar problem, "vagueness is inevitable" and here I agree with him.

8.3   Can the Biostatistical Theory Explain Dysfunction?

The main advantage of etiological theories like GSE is that they capture function's explanatory depth. Other theories, like Boorse's BST or Cummins's CR theory, do not. What about the possibility of dysfunction? Does GSE have an advantage here, too? I think it does, even if it's not as decisive as with the problem of explanatory depth. I will stick with comparing BST with GSE, since there's a fairly well-developed literature on the topic. The literature about whether or not CR can explain dysfunction is rather slim.

The claim I'll argue for is this: Boorse's BST faces three serious obstacles in accounting for function's normativity. GSE sails through all three obstacles effortlessly. I won't say BST cannot surmount those obstacles. My point is that since GSE lets us understand dysfunction in a simple way, and without the need for invoking potentially ad hoc restrictions or conceptual convolutions, it's preferable.

If you recall, according to BST, the physiological function of a trait is its species-typical contribution to fitness. (To be more specific, Boorse relativizes the function of a trait to a reference class within a species. He does this by dividing the species into subgroups on the basis of age and sex. This allows us to say things like: "The function of testosterone in a male during infancy is different from its function during puberty." I am going to ignore this nicety and just focus on species-general functions.) In the spirit of BST, the most obvious way to define "dysfunction" is in terms of statistics. We can say that if a trait token cannot do what other tokens typically do that helps fitness, it is dysfunctional.

This simple account of dysfunction is harder to get off the ground than it seems for three reasons: atypical functions, pandemic dysfunctions, and situation-specificity of function. First, the function of a trait cannot just be what it typically contributes to fitness, since some functions are atypically performed (Millikan 1989b, 285). The probability that a given sperm will fertilize an ovum is negligible, but that's still its function. Statistical normalcy is a poor guide to function and dysfunction.

Boorse can get around this problem by saying that the function of a trait is not what it typically does, since in that case, the sperm's function would be to swim around for a few months in the testes and die. Rather, its function is to do what it typically does *when it does something or other to boost fitness* (as suggested in Boorse 2002, 93). The function of sperm is to fertilize ova because that is what they typically do just on those occasions when they boost fitness. So far, so good.

This account still hits a snag. Some traits have multiple functions, one of which is commonly performed and one of which is rarely performed, such as the male urethra's job of transporting both urine and semen. Boorse's revision would imply that only the first of these is its function, since that's what it typically does when it does something useful. Karen Neander formulated this problem in unpublished correspondence with Boorse – see Boorse (2002, 93, fn. 34). Boorse could overcome this snag by saying that the function of a trait is what it nonnegligibly does, when it does something that contributes to fitness. That works well enough, but if we go that way, we have effectively abolished reference to typicality in our account of function.

GSE lets us avoid the whole tangle of atypical functions. To figure out which function a trait has, we merely need to think about why it was selected. The male urethra was selected both for transporting urine and semen, so those are two of its functions. It doesn't matter how relatively commonly or uncommonly it performs them.

Second, Boorse's view struggles to make sense of the problem of pandemic dysfunction (Neander 1991, 182). What if a disease suddenly renders everyone blind overnight? Then, the unadorned version of BST says that eyes wouldn't have the function of sight anymore. If you think this counterexample isn't worth taking seriously because it's science fiction, consider a real pandemic, like anuran decline due to UV radiation or, for that matter, the coming afflictions of climate change. There might be many pandemic dysfunctions right around the corner.

Boorse's responses to this problem have differed somewhat in different places, but the best answer he has offered is to emphasize the historical dimension of functions. (Boorse prefers to talk about "disease" rather than dysfunction, but the point carries over easily enough.) He says that when we assess a trait's "statistical-typical" contribution to fitness, we cannot just look at its current behavior. Instead, we must look at how it has helped fitness over a long slice of time that stretches from the present far back into the past (Boorse 2002, 99). Even if there's a pandemic disease that renders everyone blind tomorrow, blindness would still be atypical when we consider a long enough slice of time. This is a fine counter to the problem, though it brings Boorse's view closer, in spirit, to the etiological approaches he rejects. Again, the problem doesn't even surface for GSE. GSE is consistent with mass dysfunction because it doesn't base function on current-day performance.

The third problem is the most serious one. This is the *situation-specificity* of function. Kingma (2010) raised this problem against Boorse, and generated a minor literature. Recall that, for Boorse, a trait's function, roughly, is its species-typical contribution to fitness. We can then say a trait dysfunctions when it cannot contribute in that way. If Frankie's heart stops beating, it cannot do what hearts typically do that helps fitness, so it's dysfunctional.

Kingma notes that this simple picture is complicated by the fact that functions are situation-specific. The function of my immune system is to release antibodies into the blood, but only when there are infections to fight. A function of my medulla is to trigger the gag reflex, but only when I am asphyxiating. If I gag when I'm not asphyxiating, that's a dysfunction. Functions are relative to situations. According to Kingma, as a rule we shouldn't say that the function of X is Y but that the function of X is Y relative to situation S (as in, "the function of the gag reflex is to initiate gagging in the situation of asphyxiation"). Generalizing, BST should say this: a trait's function, relative to a situation, is its typical fitness contribution in that situation.

This is where BST implodes. Once we relativize functions to situations, all of our normal judgments of function and dysfunction are scrambled. Suppose I overdose on paracetamol. Among other things, this causes near paralysis of my stomach. Suppose my stomach does not seize up entirely, but only digests at an extremely low rate, say, a rate of gastric emptying of 50 percent over four hours, rather than one to two hours. The problem is that that reduced rate is typical for stomachs in the situation of paracetamol overdose! Strictly speaking, since the stomach is not deviating from its typical, situation-specific, contribution to fitness, there is no dysfunction either.

Now, there is a way around Kingma's dilemma. Hausman (2011, 659) says that when we assess whether or not a trait is functional, we should consider not only whether it is contributing to fitness in the typical way in the actual situation but whether it is capable of contributing to fitness in the typical way in all the other situations it might find itself in. Garson and Piccinini (2014) arrived at the same idea independently and developed it in detail. If I overdose on paracetamol, then it is true that my stomach is digesting food at the rate typical for it in that situation, but it is not capable of digesting food at the rate typical for it in a host of other situations and therefore it is dysfunctional.

This solution is promising, even if it is hard to work out all the kinks (see Kraemer 2013; Boorse 2014; Schwartz 2014; Kingma 2015 for a growing and somewhat technical literature on the issue), but it raises another problem, which Garson and Piccinini (2014, 14) note. The problem is that, according to this way of thinking about function, almost everyone's stomach is dysfunctional, all of the time. That's because there are bound to be some possible situations where my stomach can't perform as well as the average person's, even if my stomach was never designed to operate in those situations and even if it will never be in those situations. That strikes me as counterintuitive, and it seems to run counter to ordinary usage in biomedicine.

Consider motion sickness. Some people get nauseous on a rocky boat, but most do not. Suppose I'm among the minority of people who easily get motion sickness, and you aren't. That means you can digest food at a normal rate at sea (for example) and I cannot. Thus, there is at least one situation (namely, being out at sea in a rocky boat) where my stomach cannot contribute to fitness in the manner that is typical for that situation. My stomach is dysfunctional (here and now), even if I am never going to be in that situation and even if it was not designed for that situation. Most likely, your stomach is dysfunctional, too, in some respect or another.

Let me clarify: The main problem for BST is not that everyone's stomach is dysfunctional. That doesn't strike me as a terrible problem, and, besides, you might even think there's something nicely egalitarian about it. The main problem is that it seems counterintuitive that whether or not my stomach is dysfunctional now depends on how it would perform in hypothetical situations that fall outside of its normal operating conditions and that it will never actually be in.

When it comes to the problem of situation-specificity, GSE has a clear advantage. GSE not only tells us what the function of a trait is, but it also tells us which situations are appropriate for assessing dysfunction – namely, its selective environment. A stomach that can only digest at a very low rate because of paracetamol overdose is dysfunctional because if it were in its normal environment (that is, its selective environment), it wouldn't be able to perform its function. We don't have to imagine how the stomach would behave across a potentially massive range of hypothetical situations. We don't need to embrace the counterintuitive consequence that my stomach is dysfunctional just because it wouldn't perform as well as yours in some hypothetical situation it was never meant for. The question is simply: Can it perform its function in the sorts of situations it was designed to perform that function in? If not, it's dysfunctional; if so, it isn't.

In short, GSE has an advantage over BST when it comes to explaining function's normativity. The advantage is not quite as decisive as it is for the issue of explanatory depth. In that case, the selected effects theory, and its kin, are unique. I have no doubt that with enough tinkering, we can develop a version of BST that can get around all of those problems. Still, it's a notable fact that those three problems don't even arise for GSE. Why make unnecessary trouble for ourselves?

PART III

Applications

CHAPTER 9

Function Pluralism

Scientists use "function" in different ways. Sometimes, function just means *effect*. Climate change is a function of deforestation. Poor academic performance is a function of malnutrition. Sometimes, function means *useful effect*, or *benefit*. A function of soil is to provide raw materials for society's use; others include landscaping and wildlife habitat. There are functions in math, too. A mathematical function is a relation between input and output such that for each input there is only one output. None of these are functions in the ordinary biological sense of the term, and they would hardly cause confusion for scientists or philosophers.

Closer to biology, there is another sense of function that can cause a bit of trouble if we're not careful. Biologists routinely distinguish an organ's *structure* and its function. An organ's structure is how it's typically built – that is, what parts it has and how those parts are connected. The organ's function is what those parts do when they're set into motion. The function of the medial pterygoid muscle is to move the lower jaw up and forward. This is a brute mechanical fact about the muscle; you can test it by looking at it and moving it around. Here, function means something very close to effect – with this one difference – that it's intimately tied to the idea of structure. It doesn't seem tied, in any essential way, to history or evolution.

Sometimes, when biologists talk about something's function, they are using the term in this fairly minimal sense, where functions are effects that are intimately tied to structures. The biologists Bock and von Wahlert (1965, 274), for example, invoke this sense of function when they say that a trait's functions are "all physical and chemical properties arising from its form." Amundson and Lauder (1994, 454) endorse this broad usage and say it is the norm in functional anatomy. Their point is that although the selected effects theory captures one sense of function in biology, that's not the only sense, and in some disciplines it's not the primary one. Neander (2017a, 1151) recently articulated the idea of a "minimal function" and points to its utility in biology: "Minimal functions are mere doings, mere activities, which can contribute to outcomes that we might or might not be interested in explaining."

It's important to note that these functions – these minimal functions, if you will – are not the same as the functions I've been pursuing throughout this book. These minimal functions neither have explanatory depth, nor do they allow for the possibility of dysfunction, nor do they allow a distinction between function and accident. It's not as if there are two theories of function, GSE and the "minimalist" theory, which are both trying to capture the same facts about nature. They're just about two different things. All of this should be fairly unremarkable.

Today, most philosophers are happy to acknowledge multiple senses of function. They are *pluralists* about functions. Pluralism says that since biologists use "function" in different ways, there is probably more than one correct philosophical theory about it. Another way of putting it is that the word "function" means different things; different theories of function just pick out different strands of real biological usage. (That doesn't mean everyone's theory is right. It just means there can be more than one correct theory.) The only two questions that remain are, first, which senses of "function" do biologists rely on, and second, when, or in which contexts, do biologists apply the one or the other?

Although I accept pluralism, many philosophers have developed a very problematic system for bringing order into this plurality of functions. I call their system *between-discipline pluralism*, and it joins two intellectual moves (Garson forthcoming a). The first move is to say that there are only two main theories of function that biologists implicitly appeal to: the selected effects theory (SE) and the causal role theory (CR). The second move is to say that the various subdisciplines of biology, like genetics, ecology, ethology, neuroscience, anatomy, and so on, can be neatly divided up by which theory of function its practitioners normally use. Crudely put, there are SE-disciplines and CR-disciplines. For example, you might think that ethology is an SE-discipline and genetics is a CR-discipline. There are some complications I'll discuss soon, but you get the basic picture.

I think this form of pluralism is completely wrong, and it's wrong in a way that warps our understanding of how biology works. To take the first move of between-discipline pluralism: While I agree that there are at least two main strands of usage in biology, I don't think SE and CR actually capture them. First, GSE is the relevant theory for capturing one strand of ordinary biological usage, not SE. Second, there's no good reason to say that CR covers this other strand, these "minimal" functions. To take the second wing of between-discipline pluralism, even if we assume that SE and CR are the right theories – or close enough to the right theories – I don't agree that subdisciplines of biology and even psychology can be neatly divided into SE-disciplines and CR-disciplines. This is an error that stems from reading SE far too narrowly, as if it only has to do with natural selection in the usual evolutionary sense.

Here's one reason why the issue of between-discipline pluralism is so important here. Sometimes, philosophers admit that the selected effects theory is relevant to biology – but then they try, intentionally or not, to marginalize it by saying that only a very small segment of biology uses it (see below). This allows those philosophers to acknowledge its validity while quietly undermining it. We must be vigilant against these attempts to deny the selected effects theory, or theories like it, their centrality in biological thought.

In the first part of this chapter, I'll question the first move of between-discipline pluralism, which is the claim that SE and CR are the two main theories of function for biology (Section 9.1). In Section 9.2, I'll challenge the second move, the claim that different disciplines of biology can be neatly sifted into these two camps, and show why this way of dividing things up is wrong.

Function pluralism doesn't just matter for philosophers. Pluralism has crucial significance for biology, too. An excellent example comes from the ENCODE Project Consortium (ENCODE stands for the ENCyclopedia Of DNA Elements – see ENCODE Project Consortium 2012). The proponents of ENCODE made headlines a few years ago with their claim that up to 80 percent of the human genome is "functional," contradicting decades of received wisdom that most human DNA is "junk." Since then, popular books have been written about this alleged revolution in genetics (Carey 2015). Intelligent design theorists have even lauded this discovery since it bolsters their view that the genome is well crafted throughout (Wells 2011).

In response to the claims of ENCODE, philosophically astute biologists argued that the 80 percent claim is a *conceptual* mistake. It stems, they claim, from the ambiguity of "function" itself (Doolittle 2013; Graur et al. 2013). Doolittle argued that the only way one could claim that 80 percent of the genome is functional is if one conflates two senses of "function," the SE sense and the CR sense (I'll call this the "conflation hypothesis"). The proponents of ENCODE, in a recent letter, responded to these attacks by explicitly adopting function pluralism. They claimed that both SE and CR play important roles in their project: "DNA sequences identified by biochemical approaches include both SE and CR elements, and genetic variation in both has been implicated in human traits and disease susceptibility" (Kellis et al. 2014). This short letter was signed by 30 biologists leading the project. Thus, the proponents of ENCODE dodged the conflation hypothesis by embracing function pluralism. (The downfall, however, is that it renders their central claim, that 80 percent of the genome is "functional," hard to understand.) Function pluralism matters outside of philosophy, and it matters for understanding what biologists are trying to do.

9.1 Selected Effects and Causal Roles

The first wing of between-discipline pluralism is that there are two main senses of "function" in biology, the SE sense and the CR sense. Why would anyone accept that all uses of "function" can be sorted into two categories, and, more importantly, why would anyone think those are the right ones?

The argument for the relevance of SE to biology is easy enough to grasp, and I've been developing it throughout the book. Sometimes, when biologists give a function to a trait, they're trying to explain why the trait is there, like the zebra's stripes. We can call this the "*why-it's-there*" sense of function. A good philosophical theory of function should help us understand how function statements can double as causal explanations. The selected effects theory, or something like it, is the best way to make sense of this aspect of biology.

By now it should be obvious that I take exception to this way of thinking. I don't agree that SE is the best way of thinking about this strand of usage; GSE is even better. Functions are selected effects, to be sure, but "selection" in this context should be understood broadly enough to include both differential retention and differential reproduction. (Of course, one might argue that neither SE nor GSE is best way of understanding this usage. You might think that the organizational theory, or the weak etiological theory, does a better job, but I already dismissed those in Chapter 3.)

The argument for CR proceeds *via negativa*. Sometimes, when biologists give functions to traits, they don't refer to history, but since CR doesn't refer to history, CR is probably the right theory of function in those cases. This argument pops up repeatedly. Consider Godfrey-Smith (1993, 200):

Once a modified version of [the selected effects] theory is in place, the explanatory role of many function statements in fields like behavioral ecology is clear. But there remain entire realms of functional discourse, in fields such as biochemistry, developmental biology, and much of the neurosciences, which are hard to fit into this mold, as functional claims in these fields often appear to make no reference to evolution or selection. These are areas in which the attractive account of functions has always been that of Robert Cummins.

Functions in biochemistry, say, are ahistorical; CR is ahistorical and therefore CR captures functions in biochemistry.

Amundson and Lauder (1994, 463) say much the same thing:

> Several aspects of current research in functional and evolutionary morphology make crucial and ineliminable use of the concept of CR function. Anatomists often write on 'the evolution of function' in certain organs or mechanical systems, and may do so with no reference to selection or to the effects of selection... unlike SE functionalists, anatomists do not *define* a trait's function by its history. CR function is non-historically defined.

For Amundson and Lauder, the lack of reference to history or selection in large swaths of functional anatomy is good evidence for the relevance of CR in functional anatomy: "As in Cummins's account, functional anatomical analyses make no essential reference to the benefits which the analyzed capacity might have, nor to the capacity's evolutionary goal or purpose" (p. 451). They agree, however, that SE might have value in other branches.

Others have built on Amundson and Lauder's work, and now it is something like a canonical formulation: namely, that much of biology is better served by CR than SE. Griffiths (2006, 3) writes, "Amundson and Lauder and I maintain that unless anatomy, physiology, molecular biology, developmental biology, and so forth turn their attention to specifically evolutionary questions, they investigate function in the causal sense." Bouchard (2013, 93) adds, "one could say that Amundson and Lauder started by showing that physiology and functional anatomy exclusively use CR functions. Then Griffiths argued that developmental biology was another CR discipline. One of my goals is to add ecology as another functional orphan (relative to SE accounts)."

Sometimes, when people make this sort of claim, that SE is inapplicable to some parts of biology because it doesn't refer to history, they're not trying to support CR per se. Walsh (1996, 558) says, "Amundson and Lauder [1994] argue persuasively that in many subdisciplines of biology (e.g. physiology and functional anatomy) the ascription of function makes no implicit claim about history," but he uses that alleged fact to support his relational theory of function. Still, the basic idea that SE is inapplicable to large swaths of biology because biologists often don't refer to history is *de rigueur*.

Let's focus on the claim that CR captures large chunks of biology simply on account of being ahistorical. There are three big problems here. First, why think that biologists don't refer to history when they talk about functions? As we've seen, history is embedded deep inside much of basic biology, even when it doesn't appear on the surface. The concept of a creature's natural habitat is historical, and so is the idea of trait's typical contribution to fitness. (So, for that matter, is the idea of an adaptation, a sibling, a volcanic mountain, and an igneous rock.) Put somewhat differently: What exactly is it like when a biologist refers to history? Do they have to say, outright, that they are talking about history? Do they have to gather historical data for their functional hypotheses, like Caro does? Or is it enough that they deploy concepts that have an implicit historical dimension? The claim that biologists don't refer to history can't be read off surface linguistic usage, but it requires a more probing analysis.

Here is a second problem, related to the first. Can we be sure that CR is, in fact, an ahistorical theory of function – and therefore a good match for these allegedly ahistorical functions? If you recall, the biostatisical theory of function (BST) and the propensity theory are often called "ahistorical," but as the proponents of those theories acknowledged, we still find history buried deep inside of them. Why think CR is any different?

If history is embedded in CR, where exactly would it be hiding? Cummins (1975), in his canonical study, talks a lot about dispositions, analytical strategies, and hierarchically organized systems, but nothing there seems essentially historical. Even if the world came into existence five minutes ago, in exactly its present form, there would still be dispositions, analytical strategies, and hierarchically organized systems – just no deep history. His theory seems to be ahistorical through and through.

Let's take a closer look. If you recall from Chapter 8, we discovered history in Boorse's BST when we were trying to make sense of the possibility of dysfunction. First, we considered the idea that traits dysfunction just when they cannot make their typical contribution to fitness – that is, when they cannot do what the other traits do that help them survive and reproduce. Soon enough, we realized that this neat definition makes pandemic dysfunction impossible. Taking a cue from Boorse (2002, 99), we decided that the best way to account for pandemic dysfunction was to introduce history. Specifically, when we assess a trait's statistically typical contribution to fitness, we have to look at a time-slice of the species that extends far back into the past. This historical element wasn't something we were trying to read into Boorse's view; it's something he put there by himself. Maybe if CR theorists decide to give a rigorous account of what dysfunctions are, they'll feel the need to invoke history, too, for the exact same reason. The problem is that, to my knowledge, nobody has given a rigorous account of dysfunction from the CR vantage point, although some have gestured in that direction (Godfrey-Smith 1993, 200; Hardcastle 1999, 36; Craver 2001, 72).

Here is a third problem, one raised by Neander (2017a). Let's grant that biology recognizes ahistorical functions, as in the anatomists' structure-function distinction, and grant for argument's sake that CR is an ahistorical theory. Still, why assume that CR is the best account of this strand of usage? Neander (2017a, 1153) admits that there is a sense of function at play in science where functions are "mere doings, activities or causal contributions to outcomes, whether or not those outcomes are complexly achieved and whether or not anyone wants to explain them" – but she doesn't think CR captures this very well. That's because, as the passage indicates, CR adds two chunks of conceptual structure that aren't actually needed to make sense of this usage: *hierarchy* and *perspective*.

First, consider hierarchy. Recall that for CR the function of a trait is its contribution to some high-level system capacity of interest to investigators. For CR, in order for the parts of a system to have functions, they must contribute to some higher-level system capacity. Sometimes, however, functions are merely effects, even when they don't make a contribution to a well-defined system capacity. When I say that poor academic performance is a function of malnutrition, I'm not suggesting that there's a high-level system that has the capacity for malnutrition and that poor academic performance is a function of a component of that system. Maybe you can rig it up that way, but it's going to be a stretch. The same goes for the claim that a function of soil is landscaping, or the claim that the buoyant force on an object is a function of its density.

Second, consider perspective. CR functions are thoroughly perspectival. Whether or not something has a CR function, and which function it has, depends on the researcher's interests. What's perspectival about the claim that the function of the medial pterygoid is to move the lower jaw up and forward? That fact is independent of our perspectives or interests (though, trivially, it's a fact about our interests that we find it worth saying). You might try to say that the only reason I'm picking out the movement of the mandible as its function is because I have an idea that doing so contributes to chewing, which is what I care to analyze, so that function statement is implicitly relative to a functional account of chewing. It would be nice to see an argument for that claim, since I'm not convinced that there's that much implicit conceptual structure.

An irony of this whole line of thought is that Amundson and Lauder (1994, 450) – the intellectual architects of this sort of pluralism – in the midst of a hearty endorsement of the CR theory for anatomy actually admit that the anatomists' use of function isn't the same as Cummins's use; the anatomists' use is "more radical." This is because one can talk about a structure's function independent of its role in a functional analysis:

> In one way, Bock and von Wahlert's concept of function is even more radical than Cummins's. Cummins assigns functions only to those capacities of components which are actually invoked in a functional explanation, those which are believed to contribute to the higher level capacity being analyzed. Bock and von Wahlert include *all possible* capacities (causal powers) of the feature, given its current form. Some of these capacities are utilized and some are not. Both utilized and unutilized capacities are properly called functions.

For these reasons, I can't agree with the first wing of between-discipline pluralism that says SE and CR are the relevant theories of function in biology. First, it's GSE, not SE, that captures much of ordinary biological usage. Second, it's not clear that CR captures this other part of biological usage. The word "effect," (as in poor academic performance as a function of malnutrition) seems to do just fine; the word "benefit" (as in, a function of soil is to provide raw materials to humans) sometimes works, and sometimes "effect made possible by a structure" does the job, too.

9.2   SE-Disciplines and CR-Disciplines?

On to the second aspect of between-discipline pluralism. Once we have SE and CR in place as the only relevant theories for classifying all biological usage of "function," we can ask: Where and when do we apply these different theories? In other words, is there any rule for sifting biological usage and deciding which sense of "function" is at play in any given instance? Can we say anything intelligent about when, or in which contexts, scientists use SE, and when, or in which contexts, scientists use CR?

One possible answer is "no." In other words, it's possible that there are no general rules that we can easily frame about when scientists use SE and when they use CR. Maybe, in order to find out, we just have to examine each usage of "function" on a case-by-case basis. The suggestion is not that any particular instance of the term "function" is ambiguous. The suggestion is that we can, typically, decipher which sense of "function" is being used, but there are no general rules about when the one sense is used and when the other is used.

Here is a different, more optimistic, idea. Maybe the practitioners of any given subbranch of biology tend to use the term "function" in the same way. For example, perhaps the neuroscientists tend to use "function" in the CR sense, and the evolutionary biologists tend to use "function" in the SE sense. Then, whenever a scientist uses the term "function," we have a convenient rule for figuring out what that scientist means: just look up what discipline that scientist belongs to. We could even take this farther and classify various subbranches of biology into two categories, the SE-disciplines and the CR-disciplines. I call this idea "between-discipline pluralism," since it emphasizes variation between scientific disciplines when it comes to defining "function."

Many philosophers of biology have recommended just this view, as is evident from the quotations I gave in the last section. One fairly common idea here is that in the context of evolutionary biology biologists are typically using the selected effects sense of function, and in other, non-evolutionary contexts, such as ecology, neuroscience, or physiology, they are typically using CR (Godfrey-Smith 1993, 200; Amundson and Lauder 1994, 446; Griffiths 2006, 3; Maclaurin and Sterelny 2008, 114; Bouchard 2013, 86). This is a very tempting notion. After all, SE functions have to do with evolutionary origins, which is what evolutionary biologists care about. It stands to reason that if anyone uses "function" in the SE way, it's the evolutionary biologists. Everyone else must be using something in the neighborhood of CR.

There are two big qualifications. First, nobody treats between-discipline pluralism as a hard-and-fast rule. For example, it's not as if anyone thinks that all neuroscientists always use the CR notion of function, and that all evolutionary biologists always use the SE notion. All that between-discipline pluralism claims is that there's a strong correlation between which sense of "function" a scientist uses and which branch of biology he or she belongs to. Second, everyone grants that there is some overlap between scientific disciplines. If I'm a neuroscientist, but I happen to study the evolution of neurons, then I'm more inclined to use "function" in the SE sense of the term. I take it that this is what Griffiths (2006, 3) meant when he wrote: "unless anatomy, physiology, molecular biology, developmental biology, and so forth turn their attention to specifically evolutionary questions, they investigate function in the causal sense." On the one hand, he was agreeing with the basic picture of between-discipline pluralism; on the other, he acknowledged that there's some disciplinary overlap, and where there's overlap it's more likely that they will be using the SE sense.

Still, we need to resist this tempting way of dividing up functions. The problem is that the argument rests on an overly narrow conception of what SE really holds. It rests on the assumption that SE assigns functions only to traits that evolved by natural selection, but that's a mistake. As we've seen in Chapter 4, even the traditional selected effects theory applies more broadly than evolutionary natural selection; it also applies to things like antibody selection (and, more problematically, to various kinds of trial-and-error learning). Our most defensible current descendent of the selected effects theory, GSE, applies more broadly still – to anything that shows differential retention or differential reproduction in a population. This includes neural selection and (unproblematically) trial and error.

Consider the case of neural functions underlying reading ability. For example, in a recent review of the neurobiology of reading, the authors write: "we cannot yet dependably match specific brain areas to categories of function that may be impaired in a struggling decoder (e.g., visual decrowding of letters)" (Hruby and Goswami 2011, 157). Here, the authors seem prepared to say that certain brain areas have the function of "decrowding letters." That's almost certainly not an evolved function, since reading ability probably hasn't been around long enough for natural selection to have promoted it. That might lead someone to say that neither SE, nor anything in its vicinity, applies to neuroscience. These functions must be CR functions.

That conclusion doesn't follow. A bit of the brain can have a function in the GSE sense without having evolved by natural selection. It could have gotten there by neural selection instead. The same goes for the selection of learned dispositions by operant conditioning. When a rat learns to press a lever to get food, that behavior acquires a new function. One might think that GSE doesn't apply to ecology, but that would be too hasty. As I pointed out in Chapter 6, Bouchard (2013) argues that we should give functions to components of ecosystems by virtue of how they contribute to the differential persistence of ecosystems themselves (though his theory is meant to be ahistorical). If he's right, then GSE applies to ecosystems, too.

If between-discipline pluralism is wrong, then with what should we replace it? If you remember, the question I began this section with was this: Assuming that SE–CR pluralism is correct, can we say anything intelligent about when, or in which contexts, scientists appeal to SE functions, and when, or in which contexts, scientists appeal to CR functions?

I do think there is something intelligent and quite general we can say here. We can say: When scientists use "function" with explanatory and/or normative intent, they are appealing to GSE functions; otherwise, they are appealing to a different notion of function, which may or may not be captured by CR. Let me put the point more simply. When a scientist offers a function statement as an answer to a "why-it's-there" question, then he or she is appealing to GSE, in whichever discipline that statement arises. For example, suppose I ask, "Why does that rat press that lever?" One acceptable answer to the question is, "Because that lever-pressing obtains food." Another way of putting the point would be to say, "The function of the lever-pressing is to obtain food." That's a correct answer to a "why-it's-there" question.

As an alternative to between-discipline pluralism, then, I recommend *within-discipline* pluralism. Within-discipline pluralism seeks out and acknowledges the diversity of functions within any particular scientific discipline rather than between them. Within-discipline pluralism has a major advantage when it comes to scientific discovery. It encourages us keep looking for selection processes, even where we might not have initially expected to see them. It broadens our scientific horizons and helps us see new things. A good theory of function should do that, too.

CHAPTER 10

What Are Mechanisms?

The new mechanism tradition in the philosophy of science came into being about twenty years ago. Some of the ideas had been in the air long before, but they hadn't yet congealed into something like a movement. Its advocates proposed it as a new paradigm for thinking about philosophy of science. They think the notion of mechanism helps us understand the nature of scientific explanation, scientific discovery, and the history of science. As some of the core architects of that movement once wrote, "if one does not think about mechanisms, one cannot understand neurobiology and molecular biology (Machamer, Darden, and Craver 2000)."

I agree heartily that mechanisms matter to biology and science, but as I'll argue in this chapter, we get a much cleaner grasp of what mechanisms are when we see how they relate to functions. I'll make three main claims here. First, a lot of mechanism-talk in biology has a hidden, functional side; second, when we appreciate this functional side of mechanism, we can make sense of many of the features of mechanism-talk that are otherwise inexplicable; and, third, unveiling this functional side of mechanisms is vital for biomedical discovery.

Here's what I'll do in this chapter. First, I'm going to set out what I call the "functional sense of mechanism," and give three lines of evidence for it (Section 10.1). I'll then distinguish the functional sense of mechanism from another sense of mechanism, the merely *causal* sense, and set the latter aside as not relevant to my discussion (Section 10.2). Since I'm not the first to suggest that there's an intimate connection between mechanisms and functions, I'll compare and contrast my view with what others have said on the subject (Section 10.3). Then I'll develop the main argument for the functional sense of mechanism – namely that it makes sense of how mechanisms can break (Section 10.4). The functional sense of mechanism can actually help improve biomedicine, too, and I'll say why (10.5). Finally, I'll bring the functional sense of mechanism to bear on the thorny question of whether or not natural selection is a "mechanism" of evolution (10.6).

10.1 The Functional Sense of Mechanism

What are mechanisms? As I noted in Chapter 7, proponents of the new mechanism tradition have latched onto four key aspects of mechanisms: phenomenon, components (parts and activities), organization (spatial and temporal), and context (nesting and sequencing). Here's a very brief recap. First and foremost, mechanisms are identified, and individuated, by the phenomena they produce. Although we can ask about the mechanisms of demand-pull inflation, or about social cohesion in naked mole rats, or the mechanisms of skin peeling after sunburns, we can't ask how many mechanisms there are in the body, or whether the universe itself is a mechanism. That's because we haven't identified a relevant phenomenon yet.

Once we've identified the phenomenon, we can start identifying the other aspects: components, organization, context. Mechanisms are made up of components. In the standard picture, a component is an entity (e.g., an organ, a protein, a molecule) that performs an activity (beating, binding, methylating). These components are organized in a certain manner, both spatially and temporally, in such a way that they carry out certain tasks. Finally, they are inserted in a larger context, say, in a *series* of mechanisms or a *hierarchy* of mechanisms, by which they collectively carry out more complex tasks. Darden (2006, 280) gives a nice synopsis. The point here is that discovering a mechanism and filling in these details can take generations of researchers working in different fields. Hence, the new mechanism tradition doubles as a template for the history of science. The history of science is about documenting the progressive discovery of mechanisms.

For all the writing on the new mechanism tradition, philosophers in this tradition have often overlooked an extremely weird feature of biological and biomedical usage. Quite often, biologists and biomedical researchers are happy to talk about mechanisms that we associate with health and life, such as mechanisms for memory, blood circulation, digestion, and so on. When they turn to diseases, however, they often describe diseases as resulting from *breakdowns* of mechanisms for healthy states, rather than as having their own mechanisms. What is going on?

A handful of examples can prove the point. Medical researchers describe Alzheimer's disease as caused by a breakdown of a mechanism for memory: "potentially irreversible *impairments of synaptic memory mechanisms* in these brain regions are likely to precede neurodegenerative changes that are characteristic of clinical [Alzheimer's disease]" (Rowan et al. 2003, 821; emphasis mine). Addiction researchers explain drug addition in terms of the breakdown of a mechanism for behavioral reinforcement: "Only by understanding these core synaptic mechanisms can we hope to understand how drugs of abuse *usurp or modify them* (Kauer and Malenka 2007, 845; emphasis mine)." Immunologists explain autoimmune disorders in terms of the failure of immune mechanisms for recognizing self molecules: "The adaptive immune system uses multiple mechanisms to avoid damaging responses against self molecules. Occasionally, however, these mechanisms fail...causing autoimmune diseases, which can be fatal" (Alberts et al. 2012, 1539).

This is peculiar and it's worth ruminating on for a while, perhaps a long while. This split between "normal" and "broken" mechanisms suggests that, for biologists and biomedical researchers, mechanisms are implicitly tied up with functions. Any analysis of mechanism that merely talks about phenomena, components, organization, and context, misses something meaningful. Mechanisms have functions. Mechanisms are identified and distinguished by the functions they serve. Dysfunctions happen when mechanisms get disrupted, impaired, usurped, hijacked, or destroyed.

Two other lines of evidence point to this hidden, functional side of mechanisms. To see this, we can look at Stuart Glennan's (2017, 17) definition of what he calls a *minimal mechanism*: "A mechanism for a phenomenon consists of entities (or parts) whose activities and interactions are organized so as to be responsible for the phenomenon." This is meant to be an extremely broad definition that can encompass more specific ones. There is no appeal to functions here. Still, Glennan notes a peculiar consequence of his definition (p. 37). As far as minimal mechanisms go, my car is just as much a mechanism for locomotion as it is a mechanism for melting chocolate bars that are left on the seat. After all, it does both of these things and we can give a tidy mechanistic explanation for either one, in terms of components, hierarchy, and organization. Yet intuitively – at least if you and I share the same intuitions – while it seems natural and correct to say that my car is a mechanism for locomotion, it seems strange to say that it's a mechanism for melting chocolate bars. It seems equally strange to say that a car is a mechanism for spewing CO_2. But why? Nothing in the standard mechanistic account forbids that.

Darden and Craver (2013, 69) agree there's an important difference here. They capture this difference by distinguishing between a mechanism's *target* and its *byproduct*: for example, locomotion and melting chocolate bars in a car. So far, however, that just amounts to labeling the distinction. We still need an explanation for the distinction, not a label.

One obvious way to distinguish the two is in terms of function. The target/byproduct distinction maps neatly onto the function/accident distinction. The reason it seems natural to say that a car is a mechanism for locomotion, rather than for melting chocolate bars, is because locomotion is the car's function. The same point goes for spewing CO_2 into the environment. That's not the car's function; it's a byproduct. (There are parts of the car that have the function of spewing CO_2, but that's not the car's function.) Thus, the car isn't a mechanism for spewing CO_2. This confirms the bigger point: Mechanisms serve functions. In order to understand what mechanisms are, we first have to get a clear idea of what functions are.

Here is a third line of evidence that mechanisms and functions are linked. We rarely say that an organism, as a whole, is a mechanism for reproduction, or for eating, or for defecating. Yet nothing in the standard mechanistic analysis explains why not. An organism is a system organized in such a way as to result in reproduction, but it's not a mechanism for doing so. What's missing?

In my view, the answer is obvious once again: Organisms are not mechanisms for eating, or reproducing, or defecating because organisms, as wholes, don't have functions. Their parts or features have functions. My heart has the function of pumping blood, so it's a mechanism for pumping blood, but since I don't have a function, I'm not a mechanism for anything. The only exception to this rule is when we think about organisms as parts of ecosystems; in that case, it makes perfect sense to say that organisms have functions. In that context, it also makes sense to say they are mechanisms. For example, a mountain lion population can be thought of as a mechanism for reducing deer populations because that's its function in that ecosystem (e.g., Marten 2001, 35).

Let me spell out my view about mechanisms. I call it the "functional sense of mechanism" (Garson 2013):

(FSM): In order for *X* to be a (part of a) mechanism for *Y*, *X* must have the function *Y*.

This is a necessary condition; it's not a full-blown definition. One implication of my view is that if a state is dysfunctional, then there is no such thing as a mechanism for it. There's no such thing as a mechanism for cardiac arrest. There's no such thing as a mechanism for hemorrhaging.

There's no such thing as a mechanism for Alzheimer's disease. Heart disease is caused by a breakdown in a mechanism for circulation. Alzheimer's disease is caused by a disruption in a mechanism for memory. (To complicate the picture somewhat, I'm taking it for granted that all of these states represent dysfunctions. If it turned out that Alzheimer's disease did have a function, then there would be a mechanism for it. More on this soon.)

10.2 Two Senses of "Mechanism"

On the face of it, there's a serious problem for my view. Mechanisms serve functions, and functions are selected effects. That means that in order for something to be a mechanism for Y, it must have Y as its selected effect (reading "selection" broadly here to include differential retention as well as differential reproduction). This seems far too narrow. Scientists do describe mechanisms for things that weren't selected for. Glennan (2005, 445) points out that scientists talk about mechanisms for geyser eruptions, and mechanisms for El Niño phenomena, but geysers and El Niño phenomena don't have functions.

I think we can handily solve Glennan's problem by pointing out that the term "mechanism," as it's used in science, is ambiguous, just like "function" itself. So, he's right that according to one sense of "mechanism," there are mechanisms for geysers, but that's not the ordinary biological sense. Moss (2012, 165–166), does a careful job sorting out these two senses (he actually outlines three senses, but I'll leave one aside). In some contexts, when scientists ask whether or not there's a mechanism for such-and-such, all they're asking about is whether there's a plausible *causal pathway* for such-and-such. Sometimes mechanism-talk just amounts to causation-talk. Consider: "I don't believe in the efficacy of intercessory prayer, because I can't imagine a mechanism for it." When someone says there's no mechanism for intercessory prayer, he or she is just saying that there's no plausible causal pathway that leads from prayer to healing.

For another example, recall that Galileo couldn't accept Kepler's thesis that the tides are due to the moon's gravitational pull because he couldn't imagine a mechanism for it. In other words, he just couldn't imagine any plausible causal pathway from one to the other. That's the sense of mechanism at play when we ask about the "mechanisms" underlying geysers and El Niño phenomena. We just want to know what causes them. This causal sense of mechanism is irrelevant to my discussion. I want to know what mechanisms are in the ordinary biological context, as used by biologists and biomedical researchers. The main difference between the causal sense of mechanism and the ordinary biological sense is that the causal sense isn't normative – it doesn't make sense of how mechanisms can break (see Section 10.4). It makes no sense to say we didn't have a strong El Niño this year because the mechanism for it was broken. It makes no sense to say that Old Faithful failed to erupt this time around because its mechanism wasn't working right. (Granted, we sometimes talk about geological cycles getting disrupted by human activities, such as the carbon cycle. I have a hard time seeing that that talk amounts to saying that the mechanism for stabilizing atmospheric CO_2 concentrations is broken in anything more than a metaphorical sense.)

Incidentally, when Glennan formulated his influential definition of "mechanism," he was trying to explicate the very idea of causation. He wasn't concerned, first and foremost, with understanding how "mechanism" is used in biology and biomedicine; he wanted to know how "mechanism" is used throughout all of science. It's hardly surprising that Glennan doesn't think there's any intrinsic functional side to mechanisms. There isn't one, given the problem-context that was driving his work. Levy (2013) also emphasizes this point about the intellectual context of Glennan's work.

One might think that sometimes this "mere causation" sense of mechanism sneaks into biology and biomedicine, too. After all, sometimes biomedical researchers do ask about the mechanisms for dysfunctional states. They talk about mechanisms for diabetes or cancer metastasis (e.g., Himsworth 1939; Harper et al. 2016). Maybe when they say things like that, they're using mechanism in the "mere causation" sense.

However, I don't want to take that line. For, often enough, even when scientists talk about "mechanisms" for dysfunctions, they're still appealing to the functional sense of mechanism. We can see this by distinguishing between a mechanism for Y and a mechanism underlying Y. X is a (part of a) mechanism for Y if X has the function Y. X is a (part of a) mechanism underlying Z if X is a mechanism for Y and Y plays some causal role in Z. For example, in a recent article on cancer metastasis, the mechanism identified, and described as a "mechanism," was merely a mechanism for cell elasticity (Rolli et al. 2010). Cell elasticity can be a functional and adaptive quality, but can be hijacked in such a way as to facilitate the spread of cancer. Therefore, we should say that cell elasticity is a mechanism underlying cancer, but not a mechanism for cancer. Cell elasticity helps cells migrate, and cell migration is involved in cancer. Sometimes, when biomedical researchers talk about mechanisms "for" diabetes or heart disease, they're really just looking for mechanisms underlying those dysfunctions, which is consistent with my view. '

10.3   Convergence and Divergence

When we turn to discussions of mechanism in philosophy and science, we see that others have observed this weird feature of mechanisms and tried to grapple with it – although, in my view, inadequately. I'll review those attempts and say how my own view (FSM) differs.

The theoretical biologist George Williams made one of the first attempts to define "mechanism" explicitly. Intriguingly, he defines mechanism in terms of function, and function in terms of selection: "The designation of something as a means or mechanism for a certain goal or function or purpose will imply that the machinery involved was fashioned by selection for the goal attributed to it" (Williams 1966, 9). In several passages, he insists that it makes no sense to speak of "mechanisms" for things that weren't shaped by selection for their job: "Should we therefore regard the paws of a fox as a mechanism for constructing paths through snow? Clearly we should not" (p. 13). This makes Williams' view almost identical to my own. The main difference is that I wouldn't say that mechanisms need be shaped by natural selection in the evolutionary sense. Rather, according to GSE, a system can come to have a function via some other sort of selection process, not just natural selection.

Other evolutionary theorists have adopted Williams' usage. For example, evolutionary psychologists like David Buss and his colleagues often implicitly, and sometimes explicitly, restrict use of the term "psychological mechanism" to evolved dispositions that were shaped by natural selection because of their usefulness to our early ancestors (e.g., Tooby and Cosmides 2006, 185; Buss 2008, 89). Interestingly, Buss doesn't mention the work of Williams; this makes me think that this functional sense of mechanism is deeply embedded in normal biological and evolutionary usage.

Among philosophers, Lenny Moss and Carl Craver have recognized that mechanisms are tied to functions, although I think they struggled to make sense of this connection. Gualtiero Piccinini (2010, 286) also ties mechanisms to functions when he notes how "different notions of mechanism may be generated by employing different notions of function," and he points out that one could define mechanisms in terms of the etiological account of function. Alex Rosenberg (2018, 16) also says that in order for something to be a mechanism, it must have a function: "But there is one more thing that is clear about X whether it is a process the mechanism engages in or the outcome of that process. X is almost always describable as a function of the mechanism in question."

Moss (2012) outlines three different senses of "mechanism" at play in biology. The first of these senses is intertwined with purpose and goal-directedness: "To count as a biological mechanism the phenomenon in question thus must be perceived as being an expression of the ostensible 'purposiveness' of the living cell or organism" (p. 165). In short, he says X is a mechanism for Y only if X is a component, or part, of a goal-directed system.

Moss supports his contention by pointing to an interesting feature of biologists' usage. He says that biologists typically don't use the term "mechanism" when they're describing artifacts. This leads him to think there is a deep connection between mechanisms and goal-directedness in living systems:

> if cells stick to tissue culture plastic because of a chemical reaction with the plastic that resulted in the happenstance chemical production of an epoxy resin this too would be registered as an artifact and not as a 'mechanism of adhesion'...[I]t is strictly the teleological aspect which makes the difference. (ibid.)

Where I part ways with Moss is in his specific analysis. In the functional sense of mechanism, there aren't any mechanisms for dysfunctions. There are only mechanisms for functions. Moss disagrees (personal communication). He thinks there are mechanisms for diabetes, and liver damage, and schizophrenia. Where do we differ? In his view, for X to be a mechanism for Y, X need not have the function Y. X need only be a component within a system that exhibits goal-directedness and purpose. As a result, it makes sense to talk about a mechanism for diabetes because diabetes happens inside of a goal-directed system. (Another way of putting it, perhaps better, is to say that diabetes makes a difference to the goals of a goal-directed system.) I want to insist on a tighter connection between mechanism and function. I want to say that there's no such thing as a mechanism for diabetes because diabetes isn't a function of anything. It's a dysfunctional state.

Moss doesn't accept this way of putting things because he thinks it's a basic datum of biology that biologists do talk about mechanisms for dysfunctions; a good philosophical analysis should register that fact. I agree that sometimes biologists do talk that way, but my view can be reconciled with it. As I noted above, often enough, when biomedical researchers talk about "mechanisms for" dysfunctions they're actually talking about mechanisms underlying dysfunctions.

Craver (2001, 2013) also points to an intimate relationship between mechanisms and functions. To get to his view it's helpful to distinguish two different ways that mechanisms can have functions (Garson 2017d). First, a mechanism can have a function in the sense that the parts of mechanisms have functions. Others have made this claim, including Bechtel and Richardson (1993, 17) and Piccinini and Craver (2011). Second, a mechanism can have a function in the sense that the mechanism as a whole serves a function. That is the claim I'm really focusing on. In what I'm calling the "functional sense of mechanism," in order for a system to be a mechanism for something, the system has to have a function. Someone who restricts functions to parts of mechanisms doesn't have to agree with that.

Armed with this distinction, we can see that Craver (2001) and Craver (2013) actually draw quite different pictures of how mechanisms and functions hang together. The emphasis in Craver (2001) is almost entirely on the way that parts of mechanisms have functions. He refers to these functions as "role-functions." His emphasis there is quite different from mine and somewhat orthogonal. In his 2013 paper, however, he reflects on the way that mechanisms as wholes have functions. He says: "The entities and activities that are part of the mechanism are those that are relevant to that function or to the end state, the final product that the mechanism, by its very nature, ultimately produces" (p. 141). When we brush aside a lot of details, his view and my view look pretty similar: X is (part of) a mechanism for Y only if X has the function Y.

Where we part ways is on the question of what functions are. When we are talking about a mechanism *qua* whole, the function of a mechanism, for Craver, is simply any interesting system capacity. (The function of a part of a mechanism, then, is any activity that contributes in the right way to that system capacity.) The function of a mechanism *qua* whole is explicitly relative to the interests of the investigator. There's almost no restriction on what can count as a system's function. From the point of view of the physiologist, the function of the heart is to circulate blood, because that's what the physiologist cares about. From the point of view of the stethoscope manufacturer, the function of the heart is to make beating sounds, since that's what the stethoscope manufacturer cares about. I suppose that from the point of view of the forensic psychologist using the controversial polygraph, the function of the heart is to cause a spike in blood pressure in response to certain questions, since that's what the forensic psychologist cares about.

In short, Craver and I agree that mechanisms are always mechanisms for functions; but since in his view, functions are relative to observers, he can comfortably talk about mechanisms for heart disease or Alzheimer's disease (Craver 2001, 67). That's because if a group of researchers is interested in Alzheimer's, then that becomes the "function" of the system that causes it.

There are two reasons I can't go along with Craver's expansive view of mechanisms. First, as I'll shortly argue, I don't think it makes sense of the idea that mechanisms can break. Alzheimer's disease, for Craver, isn't (or isn't merely) the result of a broken memory mechanism; it has a mechanism in its own right. This view misses things about normal biological and biomedical usage. Craver might respond to this by saying that his view can make sense of how mechanisms break; a mechanism breaks just when it cannot perform its function – it's just that the relevant functions are causal role functions, not selected effects. That is fine, but it raises the problem, noted in Chapter 8, of how we're supposed to account for the possibility of dysfunction from the framework of CR. So far, nobody has given a rigorous analysis of dysfunction from the CR point of view.

Second, and more important, I think his view of mechanisms – where mechanisms serve functions, and functions are nothing more than contributions to interesting systemic effects – is undermined by biological practice itself. If it were correct, we would expect to see systematic and predictable shifts in the way that scientists assign functions to things. For example, consider pest toxicologists. Since they are interested in killing pests, such as unwanted rodents, they should assign functions to traits in a radically different way than physiologists do. They should say things like: "The rodent's lungs have the function of dispersing toxic gas to the bloodstream," or "The ant has the function of carrying boric acid to the colony." A quick perusal of pest toxicology journals shows that they don't talk that way. They talk about the functional organization of the pest's body pretty much like everyone else does (see Garson 2013, 331).

10.4 How Mechanisms Break

Mechanisms serve functions, and functions are selected effects. What good is this view? There are three benefits of recognizing this functional sense of mechanism. The first two are conceptual; the functional sense helps us understand various features of biological usage that other senses of mechanism cannot. The third is practical; it can lead to actual advances in medicine. It helps us to organize biomedical knowledge well and to explain diseases. These are big claims that I'll defend in turn.

The first benefit is conceptual. The functional sense of mechanism explains what it is for a mechanism to break. As I indicated above, biologists and biomedical researchers use a rich and colorful lexicon to describe the ways that mechanisms break. They can "break down," get "disrupted," "impaired," "hijacked," "usurped," and "coopted." One goal of medicine is to elucidate how this happens.

Philosophers of science have also recognized that mechanisms are the sorts of things that can break. Bechtel and Richardson (1993, 19), Craver (2001, 72), Glennan (2005, 448), and Darden (2006, 259) all point out that we can discover a lot about how a mechanism works by strategically breaking it in various places. Neuroscientists have exploited this strategy to great effect for nearly two centuries, ever since the pioneering work of the French biologist Pierre Flourens, who damaged pigeons' brains to figure out what the different parts do.

Here's the problem: Scientists and philosophers recognize that mechanisms can break, but standard analyses of what mechanisms are can't make sense of this simple fact. That's because, according to the standard definitions, in order for X to be a mechanism for Y, X must actually be able to do Y. For example, Glennan (2017, 17) says that the components of entities must be "organized so as to be responsible for the phenomenon." Machamer et al. (2000, 3) say that the components must be organized "such that they are productive of regular changes" [i.e., the changes that constitute the mechanism's phenomenon]. Bechtel and Abrahamsen (2005, 423) say, "A mechanism is a structure performing a function in virtue of its component parts, component operations, and their organization." This last definition sounds identical to mine because it has to do with functions but for the fact that in other writings Bechtel has identified functions with nothing more than causal roles (Bechtel and Richardson 1993, 17).

The problem with all these definitions is that, if a mechanism is not disposed to do Y, then it is not, in fact, a mechanism for Y; hence, it's not a *broken* mechanism for Y. According to these definitions, for my toilet to be a mechanism for disposing of human waste products, it actually has to be able to do that. If it can't because of a corroded plug, then on standard accounts, it's not a mechanism for disposing of waste. *Ipso facto*, it's not a broken mechanism for doing that. On standard accounts of mechanism, a broken mechanism is a logical impossibility.

The problem I'm raising for mechanisms has an exact parallel in the functions literature – namely, the normativity of function. Whatever functions are, they are the sorts of things that can dysfunction or malfunction. They can fail. Any good theory has to make sense of this. As I pointed out earlier (Chapter 8), some theories of function don't make sense of this at all. Cummins (1975, 758) once said that in order for something to have function Y, it must be disposed to do Y in the right circumstances. This theory implies that functions can't fail. The same problem, in a slightly different guise, bedevils standard accounts of mechanism. We can call this the normativity of mechanism. Mechanisms are normative in the sense that just because X is a mechanism for Y, that doesn't mean X can actually do Y.

Fortunately, the normativity of functions saves the normativity of mechanisms. A mechanism breaks just when it can't perform its function. My toilet is a mechanism for disposing of human waste products because it has the function of doing so. The great thing about functions is that they linger, even when the corresponding disposition no longer exists. Function acts like a hook on which we can hang the idea of a broken mechanism. Mechanisms are the sorts of things that can break because functions are the sorts of things that can fail.

I don't think this solution should be very controversial. We've seen philosophers appeal to the normativity of functions to save the normativity of practically everything else in biology. Teleosemantics says that the normativity of function can explain the normativity of mental representation: that is, the possibility of misrepresentation (see Chapter 12). The same goes for explaining the normativity of biological information: that is, the possibility of misinformation (see Godfrey-Smith 2000, 205; Maynard-Smith 2000, 189; Sterelny 2000, 197). Some use the normativity of function to explain the normativity of health, disease, and disorder (see Chapter 11). Finally, some think the normativity of function can save the "normativity" of biological trait categories (Neander 1991; Rosenberg and Neander 2009). Crudely put, the idea is that what makes *my* heart a member of the genetic category *heart* is that it has the function of pumping blood. This is supposed to show how biological categories are "abnormality inclusive" – a malformed heart is still a heart by virtue of having the same function, even if it doesn't look or act the same as a normal heart. I don't know whether they're right; my point is that there's a precedent for the kind of move I'm making here.

There are two ways for an opponent to respond to my claim that one needs functions to explain how mechanisms break. (I thank Stuart

Glennan for suggesting these responses to me.) First, one could bite the
bullet and just deny that mechanisms can actually break. In other words,
you could just say, "OK, my broken toilet is not, strictly speaking, a
mechanism for disposing of human waste products. It can't do that, so
it's not a mechanism for doing that. Maybe it's a mechanism for other
things, like running up my water bill by constantly leaking." I don't think
this is a good response because it runs so strongly against normal biological
and biomedical usage. Alzheimer's researchers (say) would be quite sur-
prised to hear that mechanisms can't, in fact, break. Also, we'd lose the
benefits that I'm on the verge of describing.

A second response is to acknowledge that mechanisms can break but try
to explain it in some other way. Glennan suggested that when we say a
mechanism breaks, all we mean is that it's acting in a way that's surprising
(that is, it's violating my expectations) or unusual (that is, it's doing
something it doesn't usually do). I don't think either of these moves entirely
captures the idea of a broken mechanism. Consider the midbrain's medulla.
Usually it helps us breathe; occasionally (hopefully rarely) it triggers the gag
reflex. When it does trigger the gag reflex, that might be surprising or
unusual, but it's not broken. The medulla would be broken if it couldn't
trigger the gag reflex when it should. The best way to make sense of this is
in terms of function. One function of the medulla is to trigger gagging, and
when it can't do that, it's broken. Its having this function has nothing to do
with how often it does it or what we expect it to do.

In addition to explaining how mechanisms break, which is the biggest
conceptual benefit, the functional sense of mechanism has a second
conceptual benefit that's worth pointing out. It explains the distinction
between a mechanism's target and byproduct, such as locomotion and
melting chocolate bars for cars, or pumping and making noise for hearts.
As I mentioned earlier, Craver and Darden (2013, 69) label the distinction
but don't explain it. The distinction is obvious on the functional sense of
mechanism: A mechanism's target is just its function; anything else is a
byproduct. The distinction boils down to the traditional distinction
between functions and accidents. The functional sense of mechanism gives
us a clean answer to messy conceptual questions.

10.5   How to Explain Disease

Here is a final benefit of the functional sense of mechanism. It helps make
biomedicine better because it recommends a good way of explaining

diseases. Sometimes, diseases involve dysfunctions. I assume that diabetes, cardiac arrests, and Alzheimer's disease involve dysfunctions. When we know, or have good reason to think, that a disease involves a dysfunction, we shouldn't try to find a mechanism for it. Instead, we should show how the disease is caused by a breakdown of a mechanism.

There's one big qualification here. Sometimes, diseases don't actually involve dysfunctions. Consider viral conjunctivitis, commonly known as "pink eye." (Sometimes pink eye is caused by viruses and sometimes by bacteria; I'm talking about the viral form.) It is commonly considered a disease, for example, by the World Health Organization, but there's no actual dysfunction. Pink eye happens because the conjunctiva gets inflamed. Inflammation is a functional, healthy, immune response to viruses. Nothing is actually going wrong, but it's still a disease. Seemingly, not all diseases involve dysfunctions. (Incidentally, I don't have any special explanation of what diseases are. I'm following mainstream classification by calling viral conjunctivitis a "disease.")

Why is it so helpful to think of diseases in terms of the breakdown of functional mechanisms rather than as having their own mechanisms? There are two reasons. First, it helps to synthesize the vast amount of information about diseases. Second, it might help us discover the causes of new diseases. For an example, consider neural tube defects. A crucial stage of fetal development in mammals is neural tube folding. There are many different diseases that result from neural tube dysfunctions. Diseases like anencephaly can occur when folding gets disrupted at its anterior end. Diseases like spina bifida can happen when it gets disrupted at its posterior end.

There are two different ways we might think about explaining these diseases. First, we might try to identify a mechanism for each one, complete with its spatial, temporal, hierarchical, and serial features. We first find a mechanism for anencephaly, then we find the mechanism for spina bifida, then we find one for craniorachischisis, and so on. The Moult Computational Biology Group at the University of Maryland is busy trying to explain various diseases by relying on the mechanistic framework, and Lindley Darden is working closely with that lab, and I applaud their work, but here is another way we could do it. We could try to identify, first and foremost, the mechanisms of normal neural tube formation, and then explain spina bifida, anencephaly, encephalocele, and craniorachischisis as resulting from so many different ways this mechanism can break. There aren't four mechanisms and four diseases; there is one mechanism that breaks in four ways.

I think the latter way of doing things is better for two reasons. First, it exposes a common etiology to many different illnesses and, in doing that, it helps to synthesize our growing amount of biomedical information. On the first way of speaking, we could have a complete mechanistic account of spina bifida, and another complete mechanistic account of anencephaly, without even knowing the purpose of neural tube folding. Second, the functional approach suggests new avenues to biomedical discovery. When we think about diseases in terms of functional mechanisms, it becomes second nature to reflect on the possible consequences of breaking the mechanism in new ways. "If anencephaly happens when neural tube folding is disrupted at the anterior end, what happens when it gets disrupted at the posterior end?" When we think about diseases in terms of ways that mechanisms can break, it becomes natural, psychologically speaking, to let one's mind wander down those paths.

Intriguingly, the World Health Organization also suggests that we think about diseases in terms of breakdowns of functional states. Many people are familiar with their *International Classification of Diseases (ICD)*, which is currently in its tenth edition and lists tens of thousands of diseases. Fewer people are familiar with its companion volume, *International Classification of Functioning, Disability and Health (ICF)* (2001). It is a short book (a few hundred pages) that tries to give a detailed classification of the basic functions of the human mind and body. Interestingly, the ICF explicitly discusses diseases as just so many deviations from normal function. For example, in a section on muscle tone function, they list several different ways this function can be impaired, such as in general paresis, paralysis, and Parkinson's disease (p. 98).

Other philosophers have also recommended the value of thinking about diseases in terms of broken mechanisms. Sara Moghaddam-Taaheri (2011, 608–610) says that pathologists should use information about functional mechanisms to account for the diversity of pathological states. I agree entirely with her point, although she doesn't situate it within a more general theory of what mechanisms or functions are. Karen Neander (2017b, 62) makes a similar point when she says that, as a rule, we should think about diseases as involving deviations from proper functions. However, she doesn't talk specifically about mechanisms.

Before moving on to the question of whether natural selection is a mechanism of evolution, there are four qualifications and clarifications that I want to make.

1   William Bechtel raised a problem with the idea that we should explain diseases in terms of broken mechanisms for functions. He pointed out that, at least chronologically speaking, we often know about a disease well before we know what underlying normal functions are involved (personal communication). In other words, sometimes we just have to start trying to explain a disease even if we don't know which functional mechanism is implicated, and we use what we know of diseases to shed light on functional mechanisms.

    It's true that we can identify something as a disease (e.g., insomnia) without knowing which functional mechanism is involved, and even if we don't know the function of the corresponding "normal" state (e.g., undisturbed sleep). Bechtel is making a point about the order in which scientists discover things, and I am making a point about the order in which they explain things. We often discover functional mechanisms because we first encounter a disease. Then we explain the disease in terms of the breakdown of the functional mechanism.

2   A related point is that sometimes we talk about mechanisms for things even when we don't know what the function of that very thing is. We talk about mechanisms for sleep, for example, even if we don't know what function sleep has. That is perfectly consistent with my view. When we say that there is a mechanism, M, for sleep, we are committed to the hypothesis that the function of M is to cause sleep. That hypothesis may be wrong. It might turn out that sleep has no function. Or it might turn out that M causes sleep, but that's not its function – it's just a byproduct. In that case, we should no longer say that M is a mechanism for sleep. We should just say that M causes sleep.

3   Lindley Darden emphasized to me that when we take a close look at a disease progression, we don't just see a functional mechanism breaking down. Often, a broken mechanism will initiate a disease process. Disease processes can also involve the recruitment (or, perhaps, cooption or usurpation) of functional mechanisms. For example, some diseases involve a dysregulated inflammation response, such as cystic fibrosis or Crohn's disease. Inflammation is a normal response to pathogens, but in those diseases it can disrupt health. When we actually zoom in on the details, we'll probably see a mix of breakdowns and cooptions. That's fine with me, too.

4   Not all diseases involve dysfunctions. Viral conjunctivitis is an example of a disease with no dysfunction. Rather, it's a functional, healthy immune response to a virus. In that case, we shouldn't say that viral

conjunctivitis results from the breakdown of a mechanism. Instead, we should talk about the mechanism for viral conjunctivitis. It's an adaptation; it's not a dysfunction. The same point goes for some mental disorders, as I'll describe in the next chapter. Generalized anxiety disorder (GAD), quite plausibly, is a functional, adaptive response to early exposure to stress. In that case, we shouldn't say that GAD results from a broken mechanism. Rather, it has its own mechanism. So when I say that, in general, we should explain diseases in terms of breakdowns of mechanisms, I'm specifically referring to those diseases that involve dysfunctions.

10.6   Is Natural Selection a Mechanism?

The functional sense of mechanism isn't just good for thinking about disease. It's also relevant for other problems in the philosophy of biology. One problem philosophers have struggled with is the question of whether or not natural selection is a mechanism in any of the senses relevant to the new mechanism tradition.

Certainly, biologists sometimes describe natural selection as one of the "mechanisms of evolution," along with drift, mutation, and migration. I take it as relatively uncontroversial that if we are thinking about it from what I called the "mere causation" sense, then natural selection is a mechanism of evolution. As an aside, there is another debate about whether natural selection should be seen as causing things, rather than as a statistical summary of a large number of low-level causal processes, such as births, deaths, diseases, fights, courtships, and so on (e.g., Matthen and Ariew 2002). I'm not going to enter into this latter debate. I'll just assume, for the sake of argument, that natural selection is a mechanism in this causal sense.

Now, why wouldn't natural selection be a mechanism in the sense relevant to the new mechanism tradition? Skipper and Millstein (2005, 336) argue that natural selection is not a mechanism because there's no distinct decomposition into parts, and also because it's too stochastic (p. 341). Havstad (2011) says there's too much variability in the parts. Barros (2008) defends the idea of natural selection as a mechanism by saying that it's a stochastic, rather than deterministic, mechanism, and that it operates at multiple levels (see Matthews 2016 for a recent overview and discussion).

The functional sense of mechanism suggests a different approach to all this. In my view, natural selection isn't a mechanism of evolution because it doesn't have a function. It creates traits with functions, such as zebra stripes, but doesn't have a function of its very own. For natural selection is not, itself, selected for anything. (To be sure, sometimes natural selection creates new selection processes, such as antibody selection, so antibody selection has a function.) In the sense of mechanism that I'm recommending here, natural selection isn't a mechanism of evolution. At best, it's a cause.

CHAPTER 11

What Are Mental Disorders?

A middle-aged government employee becomes convinced that the government inserted a tracking device in his rectum while he slept. A college freshman gets so paralyzed by anxiety that she fails all of her classes and drops out of school. A teenager squeezes a mouse to death for entertainment. All of these individuals have symptoms of mental disorders, including schizophrenia, panic disorder, conduct disorder, and perhaps others. What are mental disorders? What do these constellations of thinking, feeling, and acting have in common that makes them all the same kind of thing?

This question feeds into a bigger one: What are disorders more generally, mental or physical? Here, I'll focus on the more specific question of what mental disorders are. One reason to raise the question is that mental disorders are fascinating and they challenge some of our basic ideas about human nature and the mind. Philosophers of mind are rightly intrigued by mental disorders because they promise to shed light on other questions, such as the nature of thought, rationality, and consciousness (Graham 2010). Closer to the topic of this book, I want to know whether mental disorders should always be understood in terms of biological dysfunction.

Knowing what mental disorders are has practical benefits, too. Since the 1970s, American psychiatry has been rocked by a series of controversies about whether or not certain conditions are mental disorders. In the early 1970s, psychiatrists in the American Psychiatric Association (APA) engaged in heated debates about whether or not homosexuality was a mental disorder. The APA famously put the matter up for a vote; its members voted 58 percent to 37 percent to drop the diagnosis (Kirk and Kutchins 1992, 88). More recently, members of the APA have argued with each other about categories like premenstrual dysphoric disorder, gender identity disorder, and Asperger syndrome. They also argue about the proper boundaries of these categories. For example, can a person grieving the loss of a loved one be diagnosed with depression?

Knowing what mental disorders are can help to resolve just those kinds of debates. If we know what mental disorders are, we know what evidence we'd have to muster to decide whether a particular condition really is one. One reason psychiatrists in the early 1970s couldn't agree about whether homosexuality should belong in their classification system was that they couldn't agree on what kind of evidence is needed to prove that *anything* is a mental disorder. Some psychiatrists argued that it is a disorder because, they claimed, it stems from bad genes or bad parenting. They thought you must look at genetics and childhood to know whether there's a real disorder. Other psychiatrists argued that history is irrelevant. To decide whether something is a disorder, they thought, you only need to consider whether people with the alleged condition can be well adjusted and happy (see Garson 2015, chapter 8).

Psychiatrists still disagree about what sorts of evidence is needed to show something is a disorder. I once asked the chair of the *DSM-5* task force, David Kupfer, why the APA included premenstrual dysphoric disorder (PMDD) in the most recent edition of their manual. He said PMDD is included because it has a known biochemical mechanism and it can be treated with selective serotonin reuptake inhibitors. For him, that was enough. People can differ about what sorts of evidence you need to decide.

In this chapter, I'm going to try to make some progress on this debate about what mental disorders are. Specifically, I'm going to consider a very influential theory, Jerome Wakefield's harmful dysfunction account (1992; 1999a; 1999b). He thinks mental disorders always, logically speaking, involve inner dysfunctions. My conclusion will be largely negative. While Wakefield and I agree, in the main, about what functions and dysfunctions are, I don't think mental disorders always spring from dysfunctions. One of the most interesting implications of GSE is that some mental disorders are probably functional, not dysfunctional. It's worth taking on Wakefield's theory in particular because so many psychiatrists have explicitly endorsed it (see below).

Here's the plan for the chapter. I'll first introduce Wakefield's view, and position it in relation to more long-standing debates about mental disorders (Section 11.1). Then I'll introduce a common critique of Wakefield, known as the "evolutionary mismatch" problem (Section 11.2). Section 11.3 will present a novel version of the mismatch objection, which I call the *developmental mismatch* problem. Then I'll discuss how GSE further erodes the foundations of Wakefield's hybrid view (Section 11.4). If GSE is right, there are probably many more functions, and possibly mismatches, than we thought. Finally, I'll consider, and respond to, two potential objections (Section 11.5).

11.1   Mental Disorders and Biological Dysfunctions

Here's one way of approaching Wakefield's position. On the face of it, when we say that someone has a mental disorder, we're saying that the condition is undesirable, or it's unusual, or that it violates commonly held standards of what's good or proper in society. There are, however, many conditions that are undesirable, uncommon, or otherwise violate widely shared norms of what's proper but still aren't mental disorders. Being a neo-Nazi, or any other "home-grown terrorist," arguably meets all three, but most psychiatrists don't think it's a mental disorder. What's this something more that mental disorders have?

Here's an idea that emerged in the 1970s. What distinguishes a mental disorder from any socially disvalued condition is that when someone has a mental disorder, something has gone wrong inside of them. Consider the difference between someone who murders another because of a religious ideology – say, a man murders a doctor for performing abortions – and someone who kills because he believes that the victim was a disguised and malevolent alien. The second person surely has a mental disorder. That's because something went wrong inside of him, in his thought life, that caused him to act that way. The first might just be in the grips of a dangerous ideology; in the second, something in his brain isn't working the way it's supposed to.

That puts us in the position of having to spell out this idea of "there being something wrong inside of you" or there being something inside of you that "isn't working right." A tempting way to do so is in terms of function and dysfunction (Klein 1978; Spitzer and Endicott 1978). What distinguishes the second murderer from the first is that his mind, or his brain, wasn't functioning well. It was dysfunctional. The only thing that could make one think others are disguised aliens is a dysfunctional brain. This leads us to the idea that in order for something to be a mental disorder, it has to stem from an inner dysfunction. That isn't enough for being a mental disorder, but it's a necessary condition.

Wakefield articulated a theory of disorder along just these lines. He thinks, simply put, that disorders (whether mental or physical) are "harmful dysfunctions." Wakefield holds a hybrid view about disorders; it mixes facts and values. There are two aspects to his definition, harm and dysfunction. The "harm" part just means that the disorder is deemed harmful by the person's culture. That's the value part. The "dysfunction" part means that the disorder is caused by an inner mechanism that fails to perform its function. That's the fact part. Wakefield's view is initially attractive because it acknowledges some truth in both "normativist" and "naturalistic" views – that is, it sees the role of value judgments while nonetheless insisting on a solid natural foundation.

What are dysfunctions? For Wakefield, as for myself, functions are selected effects. We differ, however, in details. Wakefield often writes as if the only kind of selection process relevant for functions is natural selection in the evolutionary sense. That means, for him, for something to be a mental disorder it must stem from the failure of a mechanism to perform its evolved function. This move purports to place psychiatry on firm evolutionary footing. In Wakefield's view, psychiatrists should resolve debates about classification, in part, by considering evolutionary history. It's not surprising that many psychiatrists and psychologists with a biological orientation have explicitly endorsed Wakefield's theory (e.g., Klein 1999; Spitzer 1999; First 2007; Nesse 2007).

I find Wakefield's view problematic. When we dig into the empirical literature, we find some well-recognized mental disorders that don't necessarily involve dysfunctions. As I'll explain, there are reasons to suspect that disorders as varied as generalized anxiety disorder, antisocial personality disorder, and even the delusions associated with schizophrenia might actually have functions, just like viral conjunctivitis. Maybe they're adaptations, not dysfunctions.

I'm not saying that some mental disorders are actually adaptations. I am saying they might be. It is *logically* possible (it is not a contradiction in terms), *nomologically* possible (it is consistent with the laws of nature), and *epistemically* possible (consistent with what we know) that some mental disorders are adaptations. Since Wakefield's theory implies that it isn't logically possible for a mental disorder to be an adaptation, then, ipso facto, it's not possible in any other way. Why exclude this prospect out of hand? Although we might decide, after much more research, that mental disorders generally involve dysfunctions, nobody's in a position to say that yet. Moreover, there are promising new research programs premised on the idea that some mental disorders are adaptations, such as the Developmental Origins of Health and Disease (DOHaD) project (see Gluckman and Hanson 2006; Glover 2011). I'll return to DOHaD later in the chapter.

I think it's important to recognize that mental disorders don't always involve dysfunctions. To take it a step further, it's good to see that we actually don't have a clue what mental disorders are. When we have a friend or a loved one with a severe mental disorder, it can be extremely scary, perplexing, or sad. We want to understand what's going on because that understanding gives us a sense of mastery, even if a false one. It gives us a chance to pretend that we know what's happening and that, in principle, we know how to fix it. It's better to stop pretending that we really understand what mental disorders are. In fact, there are some mental health advocacy groups, like the Icarus Project, that don't even like using the term "mental disorder," because they think it conveys a false sense of understanding. Instead, they use the term "madness." "Madness," although quaint, is a term that doesn't pretend.

11.2   Mismatch or Dysfunction?

I'm not the first to argue that mental disorders don't necessarily involve dysfunctions. Many others have done the same. This is the basis for the "evolutionary mismatch" critique of Wakefield's view – see Lilienfeld and Marino (1995, 416; 1999, 406–407); Richters and Hinshaw (1999, 442); Woolfolk (1999, 662); Murphy and Stich (2000, 81–84). The idea is that some disorders, such as psychopathy, depression, or anxiety, are adaptations, not dysfunctions; they've just lost their value in our current environments.

Consider evolutionary psychology – specifically the version of evolutionary psychology that hit the academic scene in a big way in the early 1990s (Buss 2008; see Garson 2015, chapter 3). Proponents of this view often assume, as a working hypothesis, that many of our current-day mental capacities evolved in our shared Pleistocene past, from about two million years ago until about 10,000 years ago. (There are many different evolutionary approaches to psychology; the so-called Santa Barbara school is just one of them, and even this is a somewhat simplistic characterization of their ideas – see Confer et al. 2010 for a more up-to-date synopsis.)

Many evolutionary psychologists actually think some of our major mental disorders, like depression, the anxiety disorders, and even obsessive-compulsive disorder, might be adaptations, not dysfunctions. Nesse (2000), in a now-famous paper, reviews many different adaptationist hypotheses about depression. Brüne (2008) is a large survey of adaptationist hypotheses about mental disorders. If they're right – and that's a big if – then they wouldn't be dysfunctions. At worst, they'd be mismatches, like getting startled by a rubber snake. That's not a dysfunction; it's an adaptation. It's an example of a system performing its function perfectly well, just in the wrong environment.

This is such a pivotal point – that mismatches aren't dysfunctions – that it's worth lingering on for a moment. In Chapter 8, we decided that in order for a mechanism to be dysfunctional, it's not enough that it's in the wrong environment. Instead, there has to be something about its inner constitution that prevents it from performing its function. Remember the blindfold example: If being dysfunctional just amounts to being unable to perform one's function, then my eyes would be dysfunctional just because I'm blindfolded. To make sense of this, I urged the view that if something's dysfunctional, then, even if it were in its normal environment, it still wouldn't perform its function. I then identified the normal environment with the trait's selective environment.

We also saw that people describe a trait's function in different ways. Some describe a trait's function in terms of its proximal effects (like the heart's beating) and some in terms of its distal effects (bringing nutrients to cells). Following Neander's lead, I said that we should restrict functions to those effects that are the most proximal to the trait. I went beyond Neander by saying that proper functions are just proximal functions. It's not as if a trait has different functions, some proximal and some distal. Strictly speaking, it's wrong to say the function of the heart is to bring nutrients to cells. It is fallacy of division: attributing a property of a whole (the circulatory system) to a part.

Now, go back to the snake detector reflex. (I'm supposing, for the sake of argument, that there's a mechanism in the human brain that has the function of detecting snakes; see Isbell 2009 for an interesting discussion.) Technically, this is a response function. There is a mechanism that has the function of releasing adrenalin in response to a certain configuration of visual stimuli (an S-shape of a certain size and thickness, say). If I jump back at a life-like rubber snake, there's no dysfunction. Everything is working just as it should. (In fact, it would be dysfunctional if I didn't startle at the life-like snake.) My snake detector mechanism just isn't in its normal environment. Its normal environment is one in which there are real snakes slithering around, not a bunch of rubber toys. If you put the mechanism in its normal environment, with real snakes slithering around, it would respond just fine.

If evolutionary psychologists like Nesse are right, the mechanism that causes us to be depressed, for example, is the same sort of thing. It makes us to feel low when we're confronted with certain life situations, such as being attached to an unattainable goal. It has the function of helping us to detach from those goals, and thereby helping us to adjust to the reality of our situation. That's how it helped our Pleistocene ancestors. The problem is that, nowadays, getting depressed generally isn't a useful response to the situation one is facing. Struggling to find a job can cause depression. If the struggle is due to something you have no control over, like a recession that causes a low (but not exceedingly low) demand for a certain specialty, then it's not time to get depressed; it's time to redouble your efforts to get a job. There's a mismatch between the way someone is designed to respond, and the response that would be adaptive in the current situation, but there's no dysfunction. Being mismatched and being dysfunctional aren't merely two different ways of saying the same thing. They're mutually exclusive.

Mismatch hypotheses aren't universally accepted. Like any scientific hypothesis, they can be revised or rejected in light of new data. Murphy (2005) and Faucher and Blanchette (2011), for example, criticize mismatch hypotheses for phobias; Murphy (2005) criticizes a certain mismatch theory of depression, the social competition hypothesis. This process of criticism and revision is crucial for healthy science and it's to be unconditionally welcomed. Note, however, that even if we decide to drop one or two specific mismatch hypotheses, we're not entitled to throw out the whole research program. That's just not how research programs work (Lakatos 1970).

Some people don't take evolutionary psychology as a whole very seriously because they think that it's nothing more than "just-so stories." Gould and Lewontin (1979) famously chided their colleagues for inventing cheap and often unverifiable evolutionary explanations for various traits. Lewontin (1998) specifically took on the evolutionary psychologists, and a number of philosophers followed suit, most notably Buller (2005) and Richardson (2007). If you are one of those people, then you might as well move on to the next chapter now, because you're not going to accept Wakefield's view anyway. In other words, if you think we're rarely licensed to make any sorts of claims in this direction, you'll agree that Wakefield's proposal is practically useless for any real diagnostic purposes – you're already on my side, even if not for the same reasons.

11.3   Developmental Mismatches

In earlier work, I proposed a novel criticism of Wakefield's view, based on what I called a "developmental mismatch" (Garson 2010; 2015, chapter 8; forthcoming b). A developmental mismatch is a special sort of evolutionary mismatch. Often, natural selection doesn't confer a fixed phenotype on an organism, but bestows upon it a certain capacity to "select" between adult phenotypes during its development. This is called *developmental plasticity* and it's a kind of phenotypic plasticity (Pigliucci 2001).

There are two kinds of developmental plasticity relevant here. The first can be dubbed "switching." This takes place when a small handful of adult phenotypes is available to the developing organism, and the organism "selects" the most appropriate one. A famous example is the tiny crustacean *Daphnia*. If *Daphnia* eggs are released in waters swarming with predators, they grow a tough, helmet-shaped head. This is good for *Daphnia* because it helps it avoid getting eaten, though it makes it swim more slowly. If there are no predators around, it grows a normal, streamlined head. To keep things simple, we'll treat these two forms as being innate; we'll imagine that the genome encodes a program that says, roughly, "If predators, grow helmet; otherwise, no helmet." Chemical signals from the predator just trigger the right developmental path.

We can refer to the second kind of developmental plasticity as "imprinting." Here, there are a vast number of potential adult phenotypes available, depending on the nuances of the formative environment. In contrast to "switching," the local environment plays a more powerful role in shaping the form of the adult phenotype. An example is filial imprinting in goslings. Goslings can imprint onto pretty much any large, moving object that they encounter during the developmental window. There are potentially billions of phenotypes available to them depending on their environment. Being disposed to follow one's mother is one phenotype. Being disposed to follow a pair of boots is another. Being disposed to follow a porcupine is a third.

Developmental plasticity creates a new sort of mismatch in addition to the evolutionary kind. I'll stay with the "switching" case for the sake of illustration. Suppose an organism such as *Daphnia* is programmed to adopt phenotype P1 in environment E1, and P2 in E2. Suppose it is placed in environment E1, then it develops P1. Then, suppose the animal is moved to E2. Now, there is a mismatch between its phenotype and its environment, $<P1, E2>$, but it's not an evolved mismatch. The mismatch isn't due to a transition from the environment its distant ancestors faced to its current-day environment. The mismatch is due to the transition from its formative environment to its mature environment.

Here is the question: Is there any analogy to *Daphnia* in the realm of mental disorders? Is it possible that some mental disorders represent developmental plasticity? Consider generalized anxiety disorder (GAD). The *DSM* characterizes GAD as an excessive, diffuse, and disproportionate worry that disrupts normal life activities like work or school (American

Psychiatric Association 2013, 222). Some psychiatrists believe that GAD, and anxiety disorders more generally, result from developmental plasticity. They're not dysfunctions.

The psychiatrist Vivette Glover (2011) points out that children who grow up in high-stress formative environments are more prone to anxiety disorders as adults. This is a well-established correlation (e.g., Heim and Nemeroff 2001; McGowan et al. 2009). She goes on to speculate about the reasons for this correlation. She hypothesizes that the high-anxiety phenotype is an adaptation designed to cope with high-stress environments. The idea is that high-anxiety individuals tend to be more vigilant to potential dangers in their environments. They're more sensitive to potential danger signals around them. That would be very helpful if the level of danger in the formative environment is a good indicator of the level of danger in the mature environment. If she's right, then GAD is an example of developmental plasticity.

If GAD results from developmental plasticity, then it'd be very easy for mismatches to come about. Suppose someone is raised in a high-stress environment, and then is moved to a low-stress environment right after the developmental "window" has shut. There is now a permanent mismatch between the individual's proneness to anxiety and the likelihood that the person will actually confront a dangerous situation. It's just like the *Daphnia* with the helmet-shaped head that is moved to safe waters.

My point isn't that GAD definitely is a mismatch. It might or might not be. My point is that GAD is a mental disorder regardless of the truth or falsity of that hypothesis. As noted above, I think it's logically, nomologically, and epistemically possible that GAD is a developmental mismatch, and also a bona fide mental disorder. Wakefield thinks it's logically impossible that GAD is both a mismatch and a mental disorder. If it's a mental disorder, it's not a mismatch, and vice versa. More on this momentarily.

11.4   Generalized Selection Processes and Mental Disorders

I haven't said anything yet about how GSE can be brought to bear on this problem. I talked about the possibility of evolutionary mismatches, and then expanded this idea to include developmental mismatches, too. Still, the significance of GSE should be obvious. GSE expands the domain of entities that can possess direct proper functions, thereby increasing the likelihood of that a given condition is functional, not dysfunctional. Any sort of selection process, including operant conditioning, antibody selection, and neural selection, can create functional traits, and also potentially result in mismatches. In short, the more function-bestowing selection processes there are, the greater the probability that any given mental disorder has a function.

The psychologists John Richters and Stephen Hinshaw (1999, 442–443) gave a simple but true-to-life illustration of the point I'm after. Consider conduct disorder, marked by "a repetitive and persistent pattern of behavior in which the basic rights of others or major age-appropriate societal norms or rules are violated" (American Psychiatric Association 2013, 469). Unlike antisocial personality disorder, conduct disorder is typically diagnosed in children. These violations might include frequently starting fights, using weapons in fights such as broken bottles, being cruel to animals, breaking into homes, or theft. Suppose a young boy is raised in an abusive home (a known risk factor; ibid. 473). Suppose that, in the abusive context, aggressive and violent behavior tendencies are implicitly reinforced. For example, imagine it's a place where unexpected outbursts of violence are a great way of signaling, "keep a safe distance from me," and they come to be reinforced because they signal this. Then, those angry outbursts, and the psychological dispositions that support them, have functions. They have the function of warning would-be abusers to stay away. To be more accurate, we should say that the underlying *disposition*, which Richters and Hinshaw call a "hostile-world orientation," has a function – namely, the function of generating violent behavior on demand.

Now, suppose the child moves to a nonabusive home, where the violent behavior isn't needed anymore. The underlying disposition – this "hostile-world orientation" – keeps generating the behavior, but there are no more benefits, only drawbacks. Again, it's just like the helmet-shaped *Daphnia* in safe waters. The psychological disposition still performs its function just fine, but it's not in the normal environment for its fulfillment. It's mismatched.

Another example of a potential function, this time involving neural selection, is substance abuse (American Psychological Association 2013, 561). Scientists who study addiction believe that drugs like cocaine effectively hijack or coopt the neural mechanisms involved in the reinforcement of behavior (Kauer and Malenka 2007). Specifically, there is a tract of dopamine neurons that runs from the midbrain to the limbic system's nucleus accumbens. This is the mesolimbic dopamine tract. Any action that causes this tract to release dopamine is reinforced. Though the exact mechanisms are still a matter of controversy, it's likely that synapse selection is involved. The release of dopamine acts as a global signal that selectively strengthens just those neural connections that caused the behaviors that led to pleasure (Schultz and Dickinson 2000, 490). Sadly, this mechanism can have terrible outcomes. If you give rats the opportunity to self-stimulate mesolimbic dopamine neurons, they'll do that to the exclusion of everything else (Witten et al. 2011). This mechanism is likely involved in drug addiction, too.

Once a synapse is selectively strengthened this way, it comes to acquire a new function. It has the function of causing the behavior that led to its differential retention. If that behavior included seeking out and using drugs, then that becomes its (direct proper) function (Garson and Papineau in prep). Some of the conditions that the *DSM-5* lists are probably adaptations, not dysfunctions.

A fascinating final example of a potential function comes from the delusions associated with the paranoid subtype of schizophrenia. One somewhat adventurous theory holds that the core dysfunction involved in paranoid-type schizophrenia is a fairly low-level perceptual one. I see a coffee mug sitting next to a coffee machine and a bowl of sugar. The perception of the coffee mug is "decoupled" from the context that lets me see a meaningful connection between the three objects. I also form connections between objects in new and unusual ways: for example, the floral design on the coffee mug with the floral print of my colleague's jacket. Clearly, the fragmentation of the perceptual scene – my inability to form meaningful connections between things – is going to make it hard to navigate my normal responsibilities (Uhlhaas and Mishara 2007; also see Bovet and Parnas 1993).

In response to this perceptual disruption, I start to form delusions. These are beliefs that society considers bizarre but that might actually help me make sense of this strange new world. They strike me with the force of an epiphany or religious conversion. For example, I might form the belief that someone is conducting experiments on me, or that people are playing games with me, or that I've been appointed for a divine mission. Everything clicks into focus in a new way. If that's right, then the delusions are learned responses to the perceptual abnormalities. The delusional system of beliefs gets "fixed" over others because of its useful consequences in the way that Papineau (1984) describes (see chapter 4). It's an adaptation, like the "hostile-world orientation" invoked to explain conduct disorder. This theory of delusion, together with GSE, implies that delusions have functions; they're not dysfunctions.

Now, it must be said, consistent with Wakefield, that even if this theory of schizophrenia is right, schizophrenia still stems from a dysfunction – namely, the low-level perceptual dysfunction that skews my ability to connect perceptions correctly. The delusions would be a secondary, adaptive, response to that dysfunction, like having a fever when you're fighting the flu. The fever is part of the constellation of symptoms that make up the flu, but it's an adaptive response to the infection. The point is that to the extent that we were inclined to think that the delusions themselves were dysfunctions, we were wrong – they're adaptations. This example shows us how easy it is to be led astray when we label something a "dysfunction." It also shows us how easy it is to go wrong in how we treat schizophrenia. If the delusions represent a secondary, adaptive response to perceptual abnormalities, then it'd be useless at best – and harmful at worst – to target the delusions as if they were the core problem (Bovet and Parnas 1993, 595).

The idea that schizophrenia might represent an adaptation isn't new. Several researchers over the last sixty years have conjectured that some aspects of schizophrenia might have adaptive benefits and might be retained precisely for those benefits. One of the most famous psychoanalytic theories of schizophrenia in the 1950s was the "double-bind" theory, which was developed by Gregory Bateson and his colleagues but inspired by the work of psychoanalyst Frieda Fromm-Reichmann (Fromm-Reichmann 1948; Bateson et al. 1956). This view said that, as a child the person with schizophrenia must have been repeatedly confronted with a kind of "lose-lose" situation (typically imposed by the mother) where any coherent response would be penalized, and which forced the child to adopt radical solutions, such as delusions and incoherent speech – although I'm not endorsing this theory. Others, like the neuroscientist and psychiatrist Solomon Snyder, suggested that the delusions of schizophrenia might have some adaptive role in coping with disorganized thoughts (Snyder 1973, 66). The idea that schizophrenia or its components could be adaptive is not some fringe theory but one that mainstream psychiatrists have entertained for decades. While some of its incarnations have little going for them – like double-bind – we can't get rid of the bigger picture.

11.5   Objections and Replies

I'll consider two criticisms in closing. I said we should reject Wakefield's theory. He says mental disorders require inner dysfunctions; I say something can be a mental disorder even when there's no dysfunction. Logically speaking, Wakefield has two ways to respond to any of the particular counterexamples I raise. First, he can say that in the case I describe there's no dysfunction, but there's no mental disorder either. Second, he could say that, in the case I describe, there is indeed a mental disorder, and there's a dysfunction, too. I'll take both responses in turn.

No Dysfunction, No Disorder

Wakefield (2000, 260) has considered the possibility of evolved mismatches before. For example, he considered the idea that antisocial personality disorder might be an adaptation to ancestral environments, not a dysfunction. If that were right, he says, we should stop calling it a "disorder," because it wouldn't be one. It would represent, "normal (though problematic) variation. . .Antisocials would be different and at odds with others in the way that, say, men and woman are different and sometimes at odds, without either being disordered." I suppose he'd be inclined to describe anxiety disorders in the same way if it were shown that some anxiety disorders are adaptive responses to developmental stress.

There are two problems with this rejoinder. First, it strikes me not only as counterintuitive but as contrary to normal medical judgment. For example, the researchers who say that disorders like psychopathy and depression are adaptations still label them as "disorders," or "diseases" (e.g., Gluckman and Hanson 2006; Glover 2011). The medical professionals who think that depression, anxiety, and psychopathy are evolved adaptations, also think that they are bona fide mental disorders. Nesse (2000), for example, thinks that depression is an adaptation, but he still thinks it's a disorder. He says it's a "medical disease" (p. 15). Thus, Wakefield's view doesn't capture normal medical usage.

Wakefield sometimes uses the analogy of fever to bolster his idea that if we were to discover that depression (say) is an adaptation, we shouldn't call it a disorder (Wakefield 2000, 259). Medical professionals believed at one time that fever was a disease or disorder. When they learned that fever is an adaptation, and that it's part of the body's concerted defense mechanisms, they stopped calling it a "disorder," and they were right to do so. He thinks that if we discovered that antisocial personality disorder, for example, is an adaptation, then we should regard it just as we regard fever.

I don't find the analogy between antisocial personality disorder and fever very convincing. I agree that fever isn't a disorder, but I think that even if we discovered that antisocial personality disorder were an adaptation, we shouldn't conclude that it isn't really a disorder. There's a deep asymmetry between these two conditions. Fever has immediate and ongoing benefits for the flu sufferer; it seems to help the immune system work better, although it's not entirely clear why. Even if antisocial personality disorder did have a benefit in Pleistocene times, it doesn't have any now, as far as we know. There's an obvious reason that medical professionals think fever isn't a disorder, quite independently of how it evolved.

There is a much deeper issue here, and it touches on the question of what the point of a theory of disorder is. (This is my second critique.) Wakefield, I believe, would be prepared to say, given compelling empirical data, that antisocial personality disorder isn't a disorder, that GAD isn't a disorder, and perhaps that schizophrenia isn't a disorder either. That seems like a bad consequence for a theory of disorder. If you want to frame a definition of "mental disorder," and your definition, combined with some empirical data, implies that antisocial personality disorder, GAD, and schizophrenia aren't really disorders, then you ought at least to acknowledge how thoroughly revisionary and stipulative the project is. It shouldn't be presented as a conceptual analysis or a theoretical definition of mental disorder. Reasons should also be given for why we should accept such a revisionary and stipulative definition.

Philosophers have tacitly accepted this principle for millennia. If a preferred analysis of a certain term implies that many of the *paradigm* cases for the term's application might not be in the term's extension, then it's time to rethink that analysis (granting that to rethink it isn't to dismiss it out of hand). If I offer an analysis of what numbers are, and it turns out that 42 might not be a number, then that's a strong sign that something's wrong with my analysis. If I offer a theory of what justice is, and it turns out that knowingly framing and executing an innocent person might not be unjust, then it's likely that something's wrong with my theory. Antisocial personality disorder, schizophrenia, and GAD are paradigm mental disorders. If Wakefield's theory implies that those might not really be mental disorders, then short of some very good reason to radically revise our ontology, we should reject that theory.

A Disorder and a Dysfunction

Now I'll move on to the second part of Wakefield's reply. Wakefield could say, of some of the cases I mention, that there is a legitimate disorder. He could also say that there's an underlying dysfunction, too, and that I've therefore failed to present a solid counterexample.

Take the hypothesis that the delusions of schizophrenia represent adaptations to some low-level perceptual abnormalities (Uhlhaas and Mishara 2007). Wakefield could accept all that but still say that schizophrenia is a disorder. It's just that the dysfunction turns out to be in a different place than where we first thought it was. We might have thought, prior to this research, that schizophrenia is a disorder and the delusions represent the dysfunction. It turns out (let's suppose) that the delusion is actually an adaptation, but there is a dysfunction involved, only at a lower, perceptual level. That is entirely consistent with his theory. Wakefield would have to hope all the other examples work out in a similar way.

Wakefield has used just this line of reasoning in the conduct disorder case. He considers Richters' and Hinshaw's thought experiment, where a child is exposed to an abusive environment and develops a disposition to engage in violent outbursts but later is moved to an environment where those outbursts are useless or harmful. Wakefield is willing to entertain the hypothesis that, in this case, there is a real mental disorder – conduct disorder. Since he thinks there's a dysfunction, too, for him, it's not a counterexample.

According to Wakefield, where exactly is the dysfunction? What, specifically, has gone wrong? Wakefield (1999b, 468) says:

> if the mechanism's function is to shape personality specifically in response to the early broader environment (not the family environment, which evolutionarily is expected to be reasonably benign) to prepare for the later broader environment, then the 'accidental' setting of personality parameters by extreme (evolutionarily unexpectable) family abuse is a dysfunction.

Here's the idea. Let's suppose that in human beings there is a universal, innate mechanism, M. M is an adaptation. Its function is to gauge the "baseline" level of violence and hostility in an individual's formative environment, and then use this information to shape that individual's personality. It works according to a simple if-then rule: If there is a high level of baseline violence in my environment, then I should adopt a hostile-world orientation and behave accordingly; otherwise, I shouldn't. But the mechanism can go awry. It's designed to detect the baseline level of violence in my broader environment – not just my home but my community and natural environment. The problem is that a home environment is a limited and distorted sample of what the broader environment is like. Maybe my parents are abusive, but nobody else in my community is and there aren't any ecological threats either. M, however, doesn't "know" that the broader environment is safe; it just uses the home environment as a proxy for the broader environment, and creates a hostile-world orientation. M is dysfunctional because it triggers a hostile-world orientation when it's not supposed to.

I think Wakefield is wrong in his assessment of dysfunction. Our disagreement traces back to the issue of indeterminacy. The case I described with the hostile-world orientation is structurally identical to the case of *Daphnia*. It seems to me that if the *Daphnia* is in an environment swarming with predators, and it grows a helmet-shaped head, and is then moved to a safe environment, there's no dysfunction. Even if it's generally in safe waters, but due to a statistical fluke, there are one or two predators nearby – enough to trigger the "helmet" phenotype – there's no dysfunction either. The mechanism responsible for sampling the immediate environment and producing the corresponding phenotype has discharged its function admirably. First, I think that's intuitive, but I can go beyond intuition here. I gave reasons, in Chapter 7, for saying that proper functions are proximal functions. If we stick to the level of proximal functions in the *Daphnia* case, we can see that there is no dysfunction. Proximally described, the function of *Daphnia*'s switching mechanism can be read as a command: If hormone H is present in the immediate vicinity, in quantities Q, trigger the developmental sequence that normally results in helmet phenotype. The same point holds in the conduct disorder case. The only way one can get the result that there's a dysfunction here is if one has incorrectly latched on to a quite "distal" description of the function in question.

Nor am I contradicting Wakefield's own commitments here. Wakefield, too, seems to agree that proper functions are proximal functions. At one point, he considers Dretske's famous magnetosome case. If you remember, Dretske (1986, 26) discusses an anaerobic bacterium with an inner magnet. The magnet aligns the bacterium with magnetic north. In normal environments, that's the direction of oxygen-free water. It needs to be in oxygen-free water or it will die. What's the function of the magnetosome: to align the bacterium with magnetic north, or to move that bacterium to oxygen-free water? Dretske says it's the former because, if you hold a bar magnet over the bacterium, it'll swim up to it (and hence to the oxygen-rich water) and die. Nothing's malfunctioning; the magnetosome performed its function perfectly well.

Wakefield says he agrees with Dretske about the magnetosome. He says that if we hold a bar magnet above the bacterium, thereby leading the bacterium to deadly waters, the magnetosome isn't dysfunctional. As he puts it, "one is inclined to judge the bacterium disordered when and only when its magnetosome does not successfully orient the bacterium to prevailing north. If one holds a magnet near the bacterium to fool it. . .one does not cause a disorder" (Wakefield 1999a, 386). Wakefield should draw a similar conclusion about the case of conduct disorder.

In short, neither of Wakefield's preferred responses work out well. I've shown that it's possible (in any of the standard senses) for a mental disorder to be an adaptation, not a dysfunction. GSE supports this position by significantly expanding the range of potential functions.

The funny thing is that Wakefield's account is probably the best we have going. Perhaps it's even the "only game in town" – but that doesn't make it right. On reflection, maybe it's not such a bad thing if we don't have an analysis of mental disorder. As I said at the beginning, it's probably better to throw up our hands and admit that we don't really know what mental disorders are then to falsely believe that we do. I'm inclined to think, as Foucault (2006 [1961]) did, that the possibility of mental illness is a kind of shadow side of reason, and that reason just isn't equipped to put a familiar face on it. Maybe we should learn how to live in that unknowingness, rather than plaster it over with tidy definitions.

CHAPTER 12

A New Kind of Teleosemantics

I lean back in my chair in my office and prepare for an upcoming class. I worry about whether I'll finish the book in time. I wonder how my kids are doing in school. My thoughts are straying this way and that. What is it for a thought to be about something? How can a thought, which is inside my head, be about something outside my head? Is there a special sort of relationship between the thought and the thing? There's surely no invisible chain that links my brain to my (future) book or my children, who are on the other side of the East River.

Intentionality – the *aboutness* of thought – is both extremely familiar and deeply mysterious. It's familiar because it's woven into every aspect of thinking and wanting. It's mysterious because it doesn't correspond obviously to any known physical relationship. As Fodor (1987, 97) famously puts it:

> I suppose that sooner or later the physicists will complete the catalogue they've been compiling of the ultimate and irreducible properties of things. When they do, the likes of *spin*, *charm*, and *charge* will perhaps appear upon their list. But *aboutness* surely won't; intentionality simply doesn't go that deep.

So what is intentionality? How can we understand it in terms of more familiar physical properties? Can we understand it in this way at all?

It's worth taking a moment to flesh out the mystery. Some representations, like beliefs, have truth conditions. They can be true or false, correct or incorrect, right or wrong. Perceptual representations can have "veridicality" conditions. They can be accurate or inaccurate. While desires don't have truth conditions, they have satisfaction conditions. They can be satisfied or unsatisfied, fulfilled or unfulfilled, successful or unsuccessful. Normal physical objects, like rocks and trees, don't have truth or satisfaction conditions. They can't be true or false, accurate or inaccurate, satisfied or unsatisfied. What do you add to a physical thing, what magic has to go into it, to give it truth or satisfaction conditions?

Here's how the chapter will proceed. I'll begin by introducing teleosemantics, which is the idea that the content of a representation – what it is about – depends on biological functions (Section 12.1). I'll explore the split inside teleosemantics between those who think contents depend on the machinery that makes the representation and those who think they depend on the machinery that uses it (Section 12.2). Section 12.3 will defend the chapter's main claim: One of teleosemantics' nagging problems, the problem of novel contents, can be solved by pairing it with GSE rather than with the traditional selected effects theory. The final section (Section 12.4) will consider a somewhat different problem for teleosemantics, the problem of distal content, and defend Dretske's (1986) solution to it. I have very little new to say on this latter problem, but it's worth going back to what Dretske did because a lot of his critics miss the mark.

12.1   Meaning and Selection

Representations aren't limited to people. I take this to be a starting point for a theory of representation. Ethologists are happy to say that some animals have representations. We say vervet monkeys use different calls to represent different predators: snakes, eagles, leopards (Seyfarth, Cheney, and Marler 1980). Cognitive ethologists like Anthony Dickinson (Dickinson and Balleine 1994) think rats can represent various goals, such as lapping up some sugar water, as well as the cause-and-effect sequences needed to achieve those goals, like pulling a cord. I think frogs and toads have representations, too, or at least a precursor to representations. For example, when a fly darts into the frog's visual field, that triggers a specific burst of midbrain activity that, in turn, triggers the tongue-snap. That burst of neural activity has a meaning; it's about the fly. Maybe if we can understand how representations evolved in our phylogenetically distant ancestors, it will be easier to understand the nature of beliefs and desires in people.

Some people are suspicious about talk of representation, particularly when it comes to animal behavior (e.g., Sarkar 2013). They think it's anthropocentric at best and unscientific at worst. If you want to know how frogs catch flies, stick to the level of the complicated cellular and molecular machinery that starts with light hitting the frog's retina and ends with a tongue-snap. All talk about representations are out of place in this physicalistic picture. This stance is problematic, however, since it assumes that if representations exist, they must somehow stand over and above the neural machinery that causes the tongue-snap in response to flies. The best response to this objection, then, is to show how representations are woven into the fabric of nature; they're not ghostly presences that hover above it.

Others are suspicious about representations because they think representations clash with the *extendedness* of cognition (Chemero 2009). Once we realize that processes outside of your skull realize cognition, we won't need to explain what you do in terms of little representations getting pushed around inside your skull. As Schulz (2018) recently argued, however, there's no clash between extended cognition and representations. In fact, when we have a theory of why and how representations evolved, we might be able to say, very precisely, why organisms found it useful to externalize some of their decision-making processes, and which ones they found it useful to externalize.

It seems to me that teleosemantics is our best current account of representation. Mainstream teleosemantics joins two ideas. The first is that representations have to do with biological functions, although, as we'll see, people disagree about how exactly representations are grounded in functions. The second idea is that biological functions are selected effects. Teleosemantics grounds representation, and ultimately truth, in terms of selection processes.

Teleosemantics has at least three main benefits. First, it's naturalistic. It shows how representations are folded into the natural world. Some other theories resort to a kind of primitivism that says that although representations are indeed part of nature, one can't give a reductionist account of them (Crane 2003, 200; Burge 2010, 298). Others say that intentionality is essentially wedded to consciousness, thus denying representations to nonconscious creatures (e.g., Kriegel 2013). I find the primitivist account intellectually unsatisfying, and consciousness-based approaches chauvinistic, but I'm willing to embrace one or the other if we can't find a suitable naturalistic alternative.

A second benefit of teleosemantics, and the main one, is that it accounts for what we can call the "normativity" of representations. By normativity of representation, all I mean is that it's possible for a representation to misrepresent a state of affairs – that is, something like representational error is possible. In the most basic case, what it is for an organism to misrepresent some feature of its environment – for example, a bird to misrepresent a crocodile as a log – is for there to be a device in its brain that fails to perform its biological function. That's where the real beauty of teleosemantics lies. Teleosemantics explains the normativity of representation in terms of the normativity of function, just as we did for mechanisms, and just as Wakefield tried to do for disorders. We're just using the same gimmick, if you will, in a slightly different context. Some people have gone so far as to suggest that teleosemantics is just a theory of misrepresentation, and not a theory of content (Millikan 2004, 63).

The third benefit of teleosemantics is that it's taxonomically general. It shows how organisms which are cognitively simpler than ourselves, like rodents and anurans, can represent things, too. This strikes me as a virtue. After all, we evolved from cognitively simpler creatures, so we would expect that those cognitively simpler creatures either have representations themselves or at least have some forerunner to representations. Teleosemantics traces a path between human and nonhuman representational capacities. That strikes me as a strength in an intellectual age created by Darwin.

The question of what representations are is different from, but closely related to, the question of what biological *information* is. Genes carry information about phenotypes. Action potentials carry information about nerve stimulation. Members of the genus *Daphnia* extract information from their environments about predators and they use this information in the process of development from larvae to adult. I suspect that once we have a good theory of what representations are, it might help us to figure out what information is. Here, I'm only going to focus on representation, not information. On the face of it, representation and information are different things, for while it sounds natural to say, or wonder whether, genes carry information about phenotypes, it doesn't seem as natural to say, or wonder whether, genes represent phenotypes. (However, see Shea 2013, who argues that genes can represent the normal environmental conditions for the success of their corresponding phenotypes.)

Not everyone thinks that teleosemantics is a promising way to approach representation. Some think that natural selection is the wrong place to look for understanding how creatures can accurately represent their environments. After all, they point out, sometimes highly inaccurate detection systems can be more useful, survival-wise, than accurate ones (e.g., Godfrey-Smith 1989; Crane 2003; Burge 2010). Another line of attack points to alleged content indeterminacy problems for teleosemantics. The idea is that natural selection can't make sense of how humans can make the sorts of fine-grained discriminations that we do make. When a fly whizzes past a frog, does the frog represent the fly as a fly? Or as a bit of food? Or as a small, moving black dot? Merely pointing to natural selection doesn't answer those questions. I'll return to these objections below; however, the interested reader should consult Papineau (2017) and Neander (2017b) for more wide-ranging recent defenses.

I want to emphasize the modesty of my ambitions here. I'm not interested in a full-scale defense of teleosemantics, although I'll sketch some of the most important objections and respond to them. Rather, I want to defend the narrower claim that if one accepts teleosemantics, one ought to use GSE as the starting point rather than the traditional selected effects theory (Garson and Papineau in prep). GSE is not only more principled, but it helps get around some of the problems that arise because of its reliance on the traditional selected effects theory.

The main problem I want to focus on is the problem of evolutionarily novel contents. The issue is this: At best, teleosemantics might help us understand what it is for frogs to represent flies or for vervet monkeys to represent eagles. It doesn't help us understand, however, how animals can represent evolutionarily novel features of their environments, such as microwave ovens or celebrities (Dretske 1986, 28; Sterelny 1990, 129; Peacocke 1992, 130). After all, teleosemantics says that in order for a representation to be about something, there must be some mechanism in place that has the evolved function of making the organism respond, in the right sort of way, to that thing. However, since no mechanism has the evolved function of making organisms respond in any particular way to microwave ovens or celebrities, teleosemantics is quite limited in making sense of our representational capacities.

Before moving on, it's worth pointing out that one can make use of teleosemantics even if one doesn't think functions are selected effects. One might agree that representation is best understood in terms of biological function, but then go on to accept some alternative theory of function (Price 2001; Abrams 2005; Nanay 2013). The crucial problem with these views is the normativity of representations. GSE is our best current account of the normativity of function. We then use this to make sense of the normativity of representation. If you want to exchange GSE for some other theory of function, you should show how your preferred theory makes sense of normativity. As I discussed in Chapter 8, this isn't a trivial task.

12.2   Producers and Consumers

There are two main versions of teleosemantics on the market: producer (or "informational") teleosemantics and consumer teleosemantics. We can

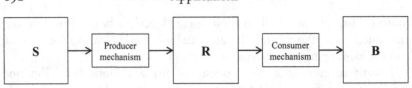

Figure 12.1   Producer and consumer mechanisms for a representation.

best understand them by setting up a simple picture (Figure 12.1). I will make the picture more complicated later.

Go back to the frog. Here, a frog is sitting on a lily pad, and a fly whizzes by. The fly triggers a signature burst of activity in its midbrain that causes the frog to snap at it. In this and other simple cases, there is some input condition S (the fly buzzing around), some output condition B (a tongue-snap), and a representation R (the burst of midbrain neural activity). In addition, there is a producer mechanism – a mechanism that causes Rs in response to Ss (which includes the optic nerve) – and there is a consumer mechanism – a mechanism that causes Bs in response to Rs (which includes bits of the hindbrain). The question, then, is this: Which side of this structure determines what the representation is about? Is it the producer side or the consumer side?

Producer teleosemantics says the content of the representation depends on the machinery that makes the representation. The best-developed version of this account is Neander's (2013; 2017b), although others who have advanced views along these lines include Fodor (1984), Dretske (1986), Jacob (1997), Schulte (2012), and Garson (2015, chapter 7). Roughly, a representation R is about something in the world, S, if and only if there is a mechanism that has the function of producing Rs in response to Ss. (This is the most basic version of the theory; we'll need to add a few bells and whistles to solve the content indeterminacy problems.) A pattern of midbrain activity is about the fly because there is a mechanism (including the optic nerve) that has the function of producing that pattern of activity in response to flies (Lettvin et al. 1959). Neander (2017b) gives a more biologically sophisticated example, using toads rather than frogs.

It's worth noting that producer teleosemantics is a theory of representational content, not representational status. The representational status question asks: What makes something a representation at all? The representational content question asks: Assuming that something's a representation, what gives it its content? It's possible for there to be a mechanism that has the function of producing Rs in response to Ss but for R to fail to be a representation at all. The medulla has the function of producing gagging in response to asphyxiation, but gagging isn't a representation of anything. Moreover, producer teleosemantics is generally pitched, in the first place, as a theory about sensory-perceptual representations – for example, what makes a bit of neural activation a representation of a blue square rather than a red triangle? It's not a theory about the whole range of representations, including concepts and beliefs. The idea is that one tries to capture a ground level of human representational capacities and then work up from there.

Neander's view relies heavily on the notion of a *response function*. Sometimes, a trait's function isn't just to make something happen but to make something happen in response to something else happening. The function of the pancreas is not just to release digestive enzymes into the small intestine but to release digestive enzymes in response to food. The function of the blood-clotting mechanism is not just to promote coagulation but to promote coagulation in response to tissue damage. Blood clotting at the wrong time and place can be fatal. Neander doesn't deny that functions are effects. The point is that sometimes functions must be finely tuned to their input conditions. In those cases, it makes sense to cite those input conditions in a full function statement: "The function of trait t is to do f in response to r." These response functions are critical to representation since the property that's represented is, more or less, the property that the mechanism has the function of responding to.

The rival to producer teleosemantics is *consumer teleosemantics*. The consumer account says that the content of the representation depends on how it's used by the system, not how it's made. Millikan (1989b) is one of the best-known advocates, but others who have advanced views along those lines include Papineau (1984; 1987; 1993), Stegmann (2009), and Shea (2007). Arguably, Shea's is a hybrid view, but since he talks of "supplementing" output-oriented teleosemantics with an input condition, I consider his view to be a modification of the standard Millikan-Papineau line.

Millikan holds, roughly, that Rs are about Ss if and only if Ss are needed for R's consumer mechanism to fulfill its function in the historically normal way (e.g., Millikan 1989b, 290). Here, the consumer mechanism involves parts of the hindbrain, which cause the tongue-snap. Its function is to help the frog eat more flies. In order to fulfill its function, the world has to be a certain way – it has to have flies around. So, midbrain neural firings are about flies. Another way of putting it: The content of the representation is the thing in the world to which the frog's motor response would be appropriate.

We should take a moment to motivate this split between the producer and consumer way of doing things. Although they typically hang together, they can come apart, and when they come apart, I think the producer side gives a more satisfying account. Consider an interesting thought experiment by Paul Pietroski (1992). He asks us to imagine a hilly region populated by two species, snorfs and kimus. Snorfs come out at sunrise and prey on kimus, but kimus can't detect snorfs. Their sensory organs just aren't responsive to snorfs and they have no evolved mechanisms to exploit natural signs of snorfs. They are completely helpless before them. (You might wonder why kimus aren't driven to extinction, but we can imagine that the snorfs, in turn, have their own natural predator that keeps snorf populations small.)

One day, a kimu is born with a mutation that produces a new neural mechanism. This mechanism causes the kimu to enter a neural state in response to certain wavelengths of light, namely, those associated with red things (that is, with a frequency in the range of 650nm). This neural mechanism also makes it attracted to red things. Every morning, when the sun rises over the hill, the kimu tries to get as close to it as possible, and it starts climbing hills. Fortunately, for the kimu, snorfs hate high places. Hence, this particular kimu manages to avoid getting eaten by snorfs. This mutation rapidly spreads through the population.

Now, after the mechanism has undergone selection, what should we say its associated neural state is about? Intuitively, it seems like we should say that the neural state is about red things, or sunrises, or wavelengths of light at around 650nm, or something along those lines. If you wanted to describe the kimu as having beliefs and desires, you'd want to say that it believes there are (say) red things nearby and that it wants to get closer to them. And that's just the content that producer teleosemantics delivers, since it says that the content of the representation is whatever in the world the perceptual mechanism is designed to respond to. (I actually think there's a certain kind of indeterminacy regarding what, precisely, its content is (see Section 12.4), but regardless of how we choose to resolve that indeterminacy, it's about something that caused the mechanism to kick into action.)

That's not the content consumer teleosemantics gives. Consumer tele-osemantics says that the neural state is about snorf-free territory, not red things. After all, for consumer teleosemantics, the representation is about whatever it has to correspond to in order for the representation's consumer mechanism to perform its function. The function of the representation's consumer mechanism is to get the kimu into snorf-free territory; to do that, the kimu's immediate environment must be snorf-free. According to Millikan's logic, we should say the representation is about snorf-free zones, not red things.

What's wrong with the claim that the neural state is about snorf-free territory? The problem is that, by hypothesis, kimus have no way of detecting snorfs. There's no mechanism in their brains that lets them divide up things in the world in terms of snorfhood. They literally wouldn't know it if a snorf were right next to them. They do have mechanisms that respond to various wavelengths of light and sort them into classes: they have *photoreceptors*. It seems sensible, therefore, that when we say that their neural states have contents, we should describe those contents in terms of wavelengths of light or in terms of something that comes before it in the chain of cause-and-effect (namely, the sun). "Snorf-free-zone" is not the right kind of content to ascribe to their neural states.

There's a more general principle here underlying Pietroski's argument. The principle is that representational content is constrained by what the organism has the ability to detect. I don't happen to have a thorough account of what it is for an organism to have the ability to detect some property. At a minimum, to say that an organism has the ability to detect a property means that there's some mechanism inside of it, and that instances of the property can, directly or indirectly, turn it on (see Ramsey 2007, 119, who explicates a corresponding notion of a "receptor").

I've made my argument for producer teleosemantics. Note, however, that very little in what I have to say requires that one accept producer teleosemantics rather than consumer teleosemantics. The key point I want to make is that GSE can benefit teleosemantics, regardless of which version we accept. Those with consumer-oriented views can benefit just as much as those with producer-oriented views. From this point forward, I'll rely on the producer view for the sake of argument, but everything I have to say about the problem of novel representations can be recast, with little modification, into the consumer approach.

12.3 Neural Selection and Novel Representations

The core claim in this section – that GSE can help us solve the problem of novel contents – comes from collaborative work with David Papineau (Garson and Papineau in prep). There, we argue that neural selection can create new representations in the brain. In that context, however, we rely on consumer teleosemantics. Here, I want to see whether the same result can be reached by using producer teleosemantics instead.

The problem, again, is this. Teleosemantics seems overly limited in its ability to account for the representational power of the mind. I'll illustrate the problem with reference to producer teleosemantics. Producer teleosemantics, in its most basic form, says that in order for a representation R to be about Ss, there must be a mechanism that was selected for producing Rs in response to Ss. At best, this theory can make sense of how people and other animals can represent evolutionary recurring features of their environments, such as snakes and spiders. It doesn't make sense of how we can represent evolutionary novel features of our environments such as flatbed auto-feed scanners or Aristotle's philosophy.

There are three ways people have tried to solve the problem of novel contents: by appeal to compositionality, derived proper functions, and ontogenetic selection processes. First, an attractive way to solve the problem is to say that novel contents are made possible by compositionality. This is a Lockean picture – that some ideas, the *complex* ones, are somehow made up of simpler ideas. Our ability to think about bear food is based on our ability to think about bears and our ability to think about food. The simple representations get their contents directly from natural selection (or some corresponding selection process) and the complex ones are somehow built out of the simpler ones. Dretske (1986, 29) and Neander (1999, 22) suggest a picture like this. Any version of teleosemantics has to give a nod to compositionality (Millikan 2004, chapter 2; see Frankland and Greene 2015 for more), but this move is limited. My idea of a flatbed auto-feed scanner can be decomposed into constituents, like flatbed, and auto-feed, and scanner, but it's not as if each of these has its own evolutionary history, and I doubt that they can be analyzed into simpler constituents, each having its own evolutionary history.

Another way of tackling the problem of novel contents is to appeal to Millikan's distinction between direct and derived proper functions. (Kingsbury 2006 gives a particularly clear statement of how this works.) We encountered this distinction earlier, in Chapter 4. We said that inside the chameleon there is a device that has a relational proper function. Its function is to make the skin match the environment. Now, suppose that the chameleon is in a brand new environment (say, aquamarine), and it manages to make the skin match that environment. The aquamarine skin, described in its specificity, has no direct proper function, since it wasn't selected for anything, but it has a derived proper function. It has the derived proper function of camouflage because it is produced by a mechanism that has a direct proper function of camouflage and that normally carries out its function by making things like that.

One can use this distinction to make sense of novel contents, too. Papineau (2006) illustrates the idea through the honeybee's waggle dance. The waggle dance represents the distance and direction of nectar. By varying certain parameters of the dance, it "says" different things about where the nectar is. Now, suppose there's some nectar in a place it's never been before, about half a mile southwest of here. Suppose the bee effortlessly represents that location by doing a dance that no bee has ever done before. That dance still means something like *there's some nectar about a half-mile southwest of here*. Because natural selection has designed a whole "system" of signs, the bee can, within its limited repertoire, "say" brand new things.

As I said in Chapter 6, I have no problem with the direct/derived proper function distinction, in and of itself. (I have a problem with using it as the sole means of giving functions to ontogenetically formed structures like configurations of synapses, but that's another story.) It can help us understand how we can represent new properties in our surroundings, but it's still quite limited. In fact, it's limited in just the way that compositionality is, since this particular solution just exploits the fact that the waggle-dance is compositional. Millikan (2004) takes pains to point out that even the simplest of signs are "architecturally complex"; they have different components, and by slightly jiggling the different components, you can "say" slightly new things. In the waggle dance, the components are the time of the dance, the place of the dance, the angle of the dance, and the speed of the dance. By changing up any of these, the bee says new things, but that strategy won't get us very far – and certainly not as far as flatbed auto-feed scanners or communism.

A third solution comes from reflecting on the diversity of selection processes in nature; it's a solution I obviously favor. Papineau develops this line of thought when he considers the idea that beliefs can undergo an ontogenetic selection process that is quite independent of evolutionary natural selection (see chapter 4; also see his 1984; 1987; 1993). Papineau (2006) elaborates this approach. He points to two different sorts of selection processes that might give rise to new functions, and hence to new contents. First, he points to nongenetic modes of inheritance (epigenetic or cultural). For an example of cultural inheritance, recall the macaque monkeys from Chapter 4 who transmit their stone-throwing proclivities behaviorally, not genetically. The point is that natural selection needn't be restricted to changes in gene frequencies but can carry on at the behavioral level as well; this gives teleosemantics new resources for explaining novel contents. Second, he points to learning processes like operant conditioning. Since operant conditioning creates new functions, perhaps it can give rise to new contents as well (I discussed this move in Chapter 4). He acknowledges that not enough detailed work has been done to bridge these alternative selection processes with the creation of new contents.

Millikan (2004, chapter 1) also recognizes that there are multiple sorts of selection processes, and that they're relevant for new contents. She recognizes at least three: the natural selection of genes, the selection of behaviors through trial-and-error, and the "mental" trial-and-error that we carry out when we're thinking about the consequences of different courses of action. (She actually considers a fourth, which is exemplified by the sort of "perceptual" trial-and-error that we make when we're thinking about how to climb up a big rock. We look at the rock from various angles and imagine what the consequences would be to tackle it from that angle. It's just like mental trial-and-error except that we're using perception in a more central way.) Now, her inclusion of mental trial-and-error seems strange, considering the view of function she gives. That's because the plans that are formed by mental trial-and-error don't undergo anything like reproduction. They represent something more like differential retention. So Millikan, in her use of examples, is already committed to something like GSE, but that commitment is obscured by her explicit doctrine of functions.

So let's begin. To show how neural selection can create new representations, consider a simple example. Suppose there are two neurons, N_1 and N_2, and both synapse onto the same target neuron N_T (See Figure 12.2.) That means there are two synapses, S_1, and S_2. Suppose that S_1 is selected over S_2 because of some effect that it has. Then we can say that S_1 now has the function of activating N_T in response to N_1. (This is a response

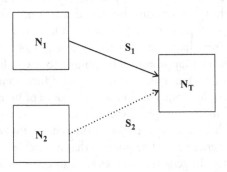

Figure 12.2   A simple case of synapse selection.

function.) According to producer teleosemantics, that is all we need for giving the representation a new content. If S_1 has the function of activating N_T in response to N_1, then the activation of N_T now represents the activation of N_1. This is the simplest possible case of representation in the brain: one neuron represents the activity of another.

Some people think we shouldn't give contents to the firing of single neurons. They think that if we say that the activity of one neuron represents the activity of another, we're somehow trivializing the very notion of representation or emptying it of explanatory labor (e.g., Sterelny 1995; Cao 2012), but this isn't a severe problem. After all, it's not as if we're attributing minds to neurons. And even if there wouldn't be much explanatory point in attributing contents to the firing of single neurons, that doesn't mean they don't have them. Sometimes the ontology that science demands runs a bit further out than its immediate explanatory needs.

If you feel strongly about this point, however, you can always add new conditions to the basic framework in order to exclude single neuron activations as representations. For example, sophisticated representations typically rely on multiple streams of informational input (Dretske 1986; Sterelny 2003; Burge 2010; Schulte 2018). Millikan (1989b, 294–297) lists six different features that human concepts possess that simpler sorts of representations don't have, such as the split between cognitive and conative representations (roughly, beliefs and desires), the capacity for negation, and so on. If you want to be a representational "chauvinist" (as Sterelny 1995 puts it), just keep adding conditions until you get something that looks like whatever you think representations ought to look like.

You might wonder how this view "scales up" above the single neuron level. In other words, I've explained how a single neuron can represent something outside of it, but how can whole clusters of neurons work together to represent something? That's no problem at all, since neural selection, in theory, can select multiple synapses at the same time. Suppose a stimulus, such as an umbrella, causes the simultaneous firing of a group of synapses. Suppose the group of synapses that responds most vigorously to the stimulus is differentially strengthened – that is, strengthened over some other group that doesn't respond quite as vigorously (see Wiggs and Martin 1998, 230; Wagner et al. 2004, 719 for discussion). When that happens, we can treat the whole group of synapses as a unit and give a function to that whole unit. We can then treat the whole group of targets as a unit for the purpose of giving it a content.

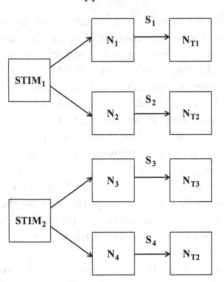

Figure 12.3   Representational capacities of sets of neurons.

I'll illustrate this through a simple example (see Figure 12.3). Suppose there are two things in the frog's environment, a fly and a worm. Suppose the fly stimulates neurons N_{T1} and N_{T2}, via N_1 and N_2 (and associated synapses S_1 and S_1). The worm stimulates neurons N_{T3} and N_{T4}, via N_3 and N_4 (and associated synapses S_3 and S_4). Suppose finally that S_1 and S_2 are strengthened over S_3 and S_4 because of some beneficial effect they produce in the presence of flies. Then it seems reasonable to treat S_1 and S_2 as a unit and assign a function to that unit. The function of S_1 and S_2 together is to stimulate N_{T1} and N_{T2}. Producer teleosemantics does all the rest. It lets us say that the simultaneous activation of N_{T1} and N_{T2} represents the fly (assuming that we can solve the problem of proximal-distal indeterminacy – which I'll discuss in the next section). There's no problem scaling the view up to make sense of how whole groups of neurons can represent things.

12.4   Proximal-Distal Content Indeterminacy

This approach to teleosemantics – effectively, a fusion of GSE and the producer account – lets us give contents to neuron firings. For example, it seems to let us say that the simultaneous firing of a group of neurons can be about a fly. Still, why say that? Why not say that the neuron is about a nutritious object? Or a small, dark, moving thing? For that matter, why say it's about something in the outside world, rather than a state of the frog, such as a specific pattern of retinal activity? These considerations lead straight to some of the trickiest problems of teleosemantics.

As noted above, Fodor's (1990) main complaint against teleosemantics has to do with various content indeterminacy problems. At the most general level, he thinks, teleosemantics can't explain how humans and other animals make the fine-grained property discriminations that they do. As it turns out, there are many different kinds of indeterminacy; Neander (2017b) discusses at least seven different problems that have been lumped under the label of "content indeterminacy." For my purposes, I'll just discuss two of them, which Godfrey-Smith (1989, 535) labels "vertical" and "horizontal."

Suppose our frog is sitting on its lily pad when a fly darts by. The fly causes a signature pattern of midbrain stimulation in the frog – call it *R*. *R* causes the frog to snap its tongue and eat the fly. We want to say that *R* represents flies: that is, that *R* constitutes something like a proto-belief of the form, *there's a fly*. (I realize that some people aren't comfortable describing the frog as having a belief; I'll call it a "proto-belief" for lack of a more convenient term. Millikan calls them "pushmi-pullyu" representations, to convey that they're neither purely indicative nor purely imperative, but both.) According to producer teleosemantics, the reason *R* means *there's a fly* is that there's a mechanism in the frog's brain that has the function of producing *R*s in response to flies.

Vertical indeterminacy stems from the fact that one and the same object of representation (e.g., the fly) has several different properties (being food, being a fly, and being a small, dark, moving object). Which one of these properties does *R* represent the fly as having? Which of these properties is *R* about? Is it, *there is a fly*? Or, *there is some food*? Or, *there is a small, dark, moving thing*? I side with Neander (2013; 2017b) on the view that if one accepts producer teleosemantics, then at the level of frogs and such, representations are generally about surface features of objects (there's a small, dark, moving thing). That's because, for producer teleosemantics, the content of *R* is whatever property it is that causally triggers the mechanism that churns out tokens of *R*. The frog's retinal neurons are causally sensitive to surface features of objects like size and shape; they're not causally sensitive to species membership or edibility.

Here, I want to focus on the problem of *horizontal* indeterminacy: Why should we say that the frog's midbrain activity represent a property of flies, rather than some more proximal stimulus, like a certain pattern of retinal stimulation? This is a very different problem. We can also call it the "proximal-distal indeterminacy problem." With the problem of vertical indeterminacy, we took it for granted that the representation was about a property of the fly. We just couldn't quite agree about which property of the fly it picks out. The problem of horizontal indeterminacy is different. It asks: What right do we have to say that the representation is about a property of some distal object, like the fly, rather than a property of my brain (say, a pattern of retinal stimulation), or even a property of some midpoint in the causal chain that extends between the distal object and my brain (say, a stream of photons coming at me)?

Before jumping into the solution, it's worth pointing out that there's an intriguing symmetry for producer and consumer accounts: Producer accounts tend to have trouble with horizontal indeterminacy, but not vertical; consumer accounts have trouble with vertical indeterminacy, but not horizontal. I'll take the second leg first. Why don't consumer accounts have trouble with horizontal indeterminacy? Because the consumer account says that the content of R is whatever it needs to correspond to in order for its consumers to do their job (Millikan 2004, 82). At first pass, in order for the consumer mechanisms to do their job, there must be an edible fly out there. If there's just a fly-shaped retinal impression, perhaps caused by a stray speck of dirt, the consumer mechanisms can't do their job. A real advantage of the consumer approach is that it seems to completely skirt this problem of distal contents.

The advantage the consumer account reaps in this area is, however, offset by a significant disadvantage in solving the vertical problem. Enç (2002, 302) calls this the "landslide argument." Suppose on the consumer view, we're trying to decide whether R means *there's a fly* or *there's an edible snack*. Then we should prefer the latter content. That's because R needs to correspond to edible snacks, not flies, for the consumer mechanism to do its job. (After all, if there's a fly buzzing around that's not edible, say, because it's toxic, then the consumers can't do their jobs.) We can follow the same line of reasoning to reach an absurd conclusion. Faced with a choice between two contents, *there's an edible snack*, and *there's something that'll help me survive*, we should prefer the latter content. After all, R needs to correspond to things that help the frog survive in order for the consumer mechanisms to do their job (if the fly is edible, but the frog can't digest it, owing, say, to an upset stomach, then the mere presence of an edible fly won't help the consumers to perform their function). That content ascription, *there's something that will help me survive*, borders on the vacuous. My claim isn't that the problem is unsolvable. My point is that both camps have their respective problems to deal with, so the mere fact that the producer approach has special problems to deal with isn't a reason to drop it in favor of the consumer approach.

With that background in mind, I can now give a solution to the horizontal problem. It helps to have a very specific scenario in mind as we work through that solution. The example I'll consider is Merriam's kangaroo rat (*Dipodomys merriami*), a rat that lives in the western United States. One of its main predators is the Mojave rattlesnake (*Crotalus scutulatus*). Scientists use this as an example of a predator–prey "arms race," where there's intense selection pressure for the prey to find new ways of evading the predator, and so constant pressure for the predator to find new ways of detecting the prey (Higham et al. 2017). Given this arms race, the kangaroo rat has a fairly sophisticated neural mechanism for detecting rattlesnakes along the lines I'll describe.

Kangaroo rats evolved in North America between five and ten million years ago. Let's assume, for the sake of argument, that the earliest rats only had one device for detecting rattlesnakes – visual information. They hadn't learned to exploit other sorts of information (smell, sound, vibrations, and so on). At the very minimum, then, there was a mechanism in their brains that responded to visual information about snakes, and produced flight behavior in response. (Kangaroo rats avoid rattlesnake strikes by leaping straight into the air.) The mechanism characterized at the crudest level would be as depicted in Figure 12.4.

Let's consider how kangaroo rats might have evolved since then. I'm taking liberties with the facts here, since we don't know all the actual details. We do know that kangaroo rats exploit visual, auditory, and tactile information to detect snakes, so my example is true-to-life even if not correct in every detail. The rat has multiple ways of getting at the same

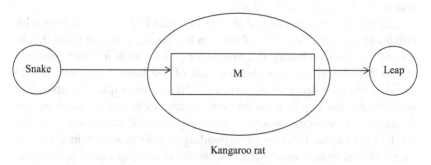

Kangaroo rat

Figure 12.4 A simple input–output mechanism.

target. Suppose a rat encounters a coiled snake. Light from the moon bounces off the snake and stimulates the rat's retina. This stream of photons constitutes a proximal, but still external, stimulus. I'll call it PS_I. (PS_I isn't the retinal activation itself but the stream of photons that are outside of the rat.) Suppose there's a group of sensory neurons that are specialized for detecting coil-shaped objects. I'll call this group of sensory neurons SN_I. When SN_I is activated, it sends a signal to a group of cortical neurons C that can also take in information from other sensory neurons. When C is activated, it triggers a group of motor neurons, M, which causes the leaping behavior. The causal chain looks like this: The rattlesnake causes PS_I, which causes SN_I-activation, which causes C-activation, which causes M-activation, which causes leaping (see Figure 12.5).

Now, we have enough neural hardware to get the producer teleoseman-ticist's story going. I'm going to assume that the activation of C is a representation, and that C-activations are about a property of rattlesnakes. (Again, there's a separate question about which property of the rattlesnake C-activations are about. That's the vertical problem. If we mix up those two problems we're doomed.) We want to know: by virtue of what are C-activations about a property of rattlesnakes? Producer teleosemantics wants to say: because there's a mechanism that has the function of producing C-activations in response to rattlesnakes. What is that mechanism? Here, I'm going to assume that that mechanism is none other than C itself. In other words, C is a mechanism that has the function of entering the "on" state in response to the rattlesnake's shape.

Here is where horizontal indeterminacy raises its head. What makes C about snakes, rather than a more proximal cause – namely, SN_I-activation? Or, a bit further out, PS_I? It doesn't seem that we have any warrant for saying that C-activations are about rattlesnakes, or that they are only about rattlesnakes. Maybe they are about rattlesnakes, but they are also about SN_I-activation and PS_I.

The problem is complicated by the fact that I identify functions with *proximal* functions. Recall my discussion in Chapter 7, where I said that in general we should identify the function of a trait with its most proximal effect. We should say that the function of the heart is to beat, not to circulate blood or bring nutrients to cells. In the last chapter, I emphasized that this also has implications for how we describe response functions. If a device has a response function, then the state of the world that the mechanism has the function of responding to is its most proximal normal cause. For example, consider the mechanism of developmental plasticity in *Daphnia*. What's its function? I said that its function is just to trigger a

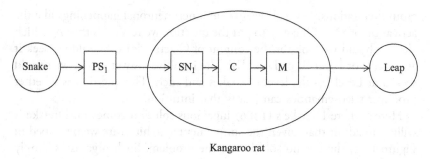

Figure 12.5   A snake-detection mechanism that can only exploit information from a single modality (vision).

certain developmental sequence (the one that normally results in helmet-shaped heads) in response to the presence of hormone H, regardless of whether there are predators around. Very precisely, when we describe the function of mechanism C in the rat's brain (as outlined in Figure 12.5) we should say that the function of C is to enter an activated state in response to SN_1-activation.

You might think that, since all my functions are proximal functions, I can't get distal contents out of them. In other words, since I think the function of C is to become activated in response to SN_1-activation, I have no basis for saying that C-activation is about rattlesnakes. I don't think the connection between functions and contents must be quite so tight. Contents should be constrained by functions, but the content of a representation needn't be identical to its function. For example, I don't think that if the function of a trait is to produce Rs in response to Ss, then Rs must be about Ss, or only about Ss. The content can be a distal cause of S. Specifically, the content will be among what Neander (2017b, 136) calls the "normal causes" of R – that is, one of the causes that plays into the selectionist account of why the producer mechanism exists. Still, even if we think that the connection between functions and contents is a loose one, we still have a large number of plausible contents to choose from. Are C-activations about snakes? Or about photons? Or about retinal activations? Or something else?

One helpful way of framing the horizontal indeterminacy problem is in terms of levels. We can imagine breaking up the chain of events depicted in Figure 12.5 into three "levels." Level I is composed of entities and processes in the outside world, such as snakes. Level II is composed of more proximal external stimuli like barrages of photons, sound waves, or ground vibrations. Level III events are inner, neuronal happenings like the activation of SN_1. One way to put the question we're after is this: At which "level" should we look for the content of C-activations? Should we seek it at Level I, or Level II, or Level III? The intuitive answer is that we should seek it at Level I – the level of snakes and such. The question is whether producer teleosemantics can justify that intuition.

Here's where Dretske's (1986) ingenious solution comes in. Dretske is willing to admit that, given the simple neural architecture we described in Figure 12.5, there is no solution to the problem. Such organisms simply can't have mental representations with determinate contents regarding snakes. As a consequence, we have no basis for saying that such organisms are capable of misrepresenting their environments (p. 32).

Suppose, however, that over some evolutionary time, kangaroo rats evolve a second system for detecting snakes. Suppose, in addition to their ability to exploit visual information, they also begin to exploit auditory information. Suppose that rats that could exploit both visual and auditory information regarding snakes were able to detect them more accurately (have fewer false negatives) than rats that could only exploit visual information. Now, we have two quite distinct groups of sensory neurons, SN_1 and SN_2, both of which feed into C, as in Figure 12.6. Otherwise, the neural architecture is unchanged.

Now, in this lineage of rats, what does any particular C-activation mean? On first glance, this new architecture seems to rule out certain contents that the old architecture let in. For we can't say that the activation of C is about SN_1-activations. That would be arbitrary, since C-activations are just as much about SN_2-activations as they are about SN_1-activations. Nor can we say that C-activations are about SN_2-activations, for the same kind of

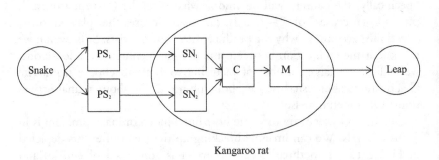

Kangaroo rat

Figure 12.6 A snake-detection mechanism that can exploit information from two different modalities (vision and hearing).

reason. (Nor can we say that C-activations are about PS_1, since they're just as much about PS_2.). We have now excluded all of the plausible Level II and Level III content ascriptions. The only plausible content ascriptions left are Level I ascriptions (i.e., the snake). (Again, we still have the problem of vertical indeterminacy to deal with. Which property of snakes are C-activations about? But that's a separate issue. For now, we just need to get contents out there, in the world, at Level I.) As long as the rat can exploit multiple sources of sensory input, then we have principled reasons for identifying the content of C-activations exclusively with snakes.

If only it were that simple! Here's a new problem that Dretske raised. Why can't we identify the content of C-activation with a disjunctive property? Perhaps we should say that in addition to being about snakes the C-activation represents the disjunctive property of being an SN_1-activation or an SN_2-activation. Additionally, perhaps we can say that C-activation represents the disjunctive property of being PS_1 or PS_2. How can we rule out these disjunctive properties?

Dretske admitted that we have no principled basis for excluding such disjunctive contents, at least when it comes to fairly simple minds. Consider, however, a slightly more complex mind that is capable of associative learning. Suppose we have a rat that, in addition to being able to respond to SN_1 and SN_2, can also associate new sorts of sensory stimuli with snakes. It can learn, during its life, to exploit new proximal stimuli as signs for snakes. Perhaps it has the capacity to learn to associate the presence of snakes with olfactory or tactile information (say, a specific odor or pattern of ground vibration).

Dretske doesn't describe this mechanism in any real detail. Let me take a moment to set out how this mechanism might work. I'll call this mechanism RECRUITER. RECRUITER's job is to help the rat associate new proximal stimuli with the snake. It's a very simple device, mechanically speaking. We can imagine that it works in the following way (see Figure 12.7). The activation of C is strongly correlated with snakes. Whenever C is activated, RECRUITER seeks out other groups of sensory neurons that are simultaneously active with C. (It's just a simultaneity detector.) Once it finds a group of sensory neurons the activation of which is correlated with C-activation, it builds a new synapse between that group of neurons and C. It simply recruits new clusters of sensory neurons, the SN_i, to act as triggers for C, whenever it's likely that SN_i is correlated with snakes.

Now, here is my question. Let's consider the lineage of rats in which this new associative mechanism evolved. What is the function of C in these rats? The function of C is to enter the activated state in response to

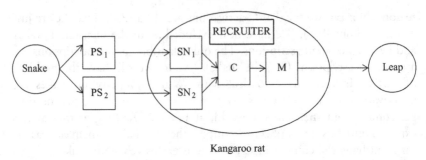

Figure 12.7   A kangaroo rat capable of associative learning.

whatever RECRUITER recruits. It's impossible to identify the content of C-activation with any disjunctive property composed of a finite number of disjuncts at Level II or Level III. We can't say that C-activation is about the disjunctive property of being an SN_1-activation or SN_2-activation. Any attempt to identify the content of C with a disjunctive property composed of a finite number of disjuncts at Level II or Level III would not accurately mirror C's function. Therefore, we do have a principled reason to exclude Level II and Level III events when we're searching for the content of C-activations.

With this picture in mind, we can respond to a criticism that Barry Loewer (1987, 306) raised against Dretske's solution. The idea is this, if I understand it correctly. If we consider any particular kangaroo rat, over its entire life, there's going to be a finite, and relatively small, set of sensory stimuli that it learns to associate with snakes. Suppose that this kangaroo rat is enjoying the last five minutes of its life. Over its long life, it has learned to exploit information about snakes from three different modalities: vision, hearing, and touch. Accordingly, whenever C has been activated throughout its entire life, it has only been activated by one of three groups of sensory neurons: SN_1, SN_2, and SN_3. Isn't it correct to say, for this rat, in its last moments of life, that the content of C-activation is the disjunctive property of being an SN_1-activation or being an SN_2-activation or being an SN_3-activation?

I don't think that we're entitled to say that – even in the given situation. Again, the whole point of teleosemantics is that content ascriptions are based on corresponding function ascriptions. That doesn't mean that the connection between functions and contents has to be airtight. Content ascriptions need to be constrained appropriately by function ascriptions (otherwise, you're not doing teleosemantics anymore). It seems to me that

Loewer's proposed content ascription, where the content of C-activation is the disjunctive property of being an SN_1-activation or being an SN_2-activation or being an SN_3-activation, just falsifies the corresponding function ascription. C's function, in this lineage, is just to enter the activated state in response to whatever RECRUITER recruits. That's an inherently open-ended ascription. Loewer's contents don't line up with the open-endedness of Dretske's functions.

Dennett (1987, 306) also criticizes Dretske's solution. He thinks that Dretske's solution, which places so much emphasis on associative learning, doesn't allow the organism to have any innate representations, and he finds that implausible. But I don't think Dretske's solution says that. First, Dretske's solution might imply, at the very worst, that there are no highly determinate innate representations. There would only be indeterminate ones – but I don't think Dretske's solution even implies that because as soon as a rat possesses RECRUITER, it need not have actually utilized RECRUITER in order to have determinate contents. Once the rat merely possesses RECRUITER (by, say, a few weeks after birth), then, at that moment, C's function is to become activated in response to whatever RECRUITER recruits. At that moment, we can no longer identify the content of C-activation with any disjunctive set of Level II or Level III happenings. If we tried to identify the content of C-activations with some disjunctive property, we'd do an injustice to C's function. The rat doesn't have to take months or years to learn about snakes before it can have determinate representations about them.

Neander (2017b) develops a third problem for Dretske's view. She agrees that Dretske's solution can probably help us understand how creatures can represent properties of distal objects. In other words, if C can be activated by many different sensory modalities (sight, sound, touch), then C-activations are about a distal object, and not states of the organism. Consider, however, a representation that is restricted to a single sense modality. For example, consider the rat's representation of the snake's shape, R_S. We want to be able to say that that representation, R_S, is about a property of the snake, not a property of the retina. Dretske's solution is silent about how representations that are restricted to single sensory pathways can have distal contents: "this strategy will not confer determinately distal contents on the sensory-perceptual representations that belong to the specific epistemic routes that we are in effect abstracting from" (p. 227).

The argument is well put, but we can tweak his solution to get the right answer. The reason that R_S is about a property of the snake, rather than a property of the retina, is because it's one small component of a more complex representation R which itself is about a distal content. R_S derives its distality from the distality of R. The point is that if Dretske's organism only had a single sensory modality (say, a worm with a cluster of photo-receptors near the surface of its skin), then it wouldn't have determinate contents at all. That sounds right to me.

Note that the same conclusion should be drawn in the fictional kimu case. Suppose our kimu only has a single sensory receptor for detecting red things. Then it seems to me that it's not capable of highly determin-ate representations. In other words, there's no principled basis for saying that the activation of that receptor means *there's the sun*, rather than *there's some wavelengths of light at around 650nm*, or *there's a red thing*. (Just because it's not capable of highly determinate representations, I still don't want to say that the content might be *there's some snorf-free territory*. To the extent that there's any indeterminacy about the content of the kimu's mental state, it's indeterminate with respect to this proximal–distal axis. The list of candidate contents should only include those items that have a causal role in triggering the perceptual mechan-ism; snorf-free zone isn't on that list.)

Schulte (2018) develops a solution to distal content that relies heavily on *constancy mechanisms*, in addition to other principles. A constancy mechanism is a mechanism that lets us represent a constant property of an object in spite of wildly fluctuating sensory input. Consider size constancy: As an object moves closer and further away from us (say, a person is running up and down a field, toward and away from us), it appears to have pretty much the same size, even though the retinal image that it projects is shifting around wildly. Or consider brightness con-stancy: The perceived brightness of a surface can seem more or less the same even when there are large changes in illumination. A white piece of paper still looks white to me whether I'm outside in bright sunlight or inside a faintly lit room. Schulte thinks that in order for a representation to have a distal content, it has to be produced by a constancy mechanism. He adds two new principles, which he calls a "naturalness" and "imme-diacy" principle, to round out his theory. Dretske (1981, 162), Sterelny (1990, 138), and Burge (2010, 408) also appeal to constancy mechan-isms to deal with distal content; Schulte's approach differs in that it adds extra principles and situates the solution firmly in Neander's producer teleosemantics.

From what I can tell, however, a core part of Schulte's solution, the part about constancy mechanisms, is just a more specific version of Dretske's solution – which he seems to acknowledge (fn. 29). A constancy mechanism, from what I can gather, is just one subtype of the more general mechanism that Dretske describes. At its most general level, the mechanism Dretske describes is one that takes, as input, many different types of sensory data, and churns out tokens of the same representation type. Constancy mechanisms are simply one way to implement this general function. The main difference is that Schulte drops Dretske's requirement for an association mechanism. That seems like a step backward, since the whole point of the association mechanism was to solve the disjunction problem. Instead, Schulte says that the way to solve the disjunction problem is to introduce a *naturalness* condition, where the most natural cause of a representation is the least disjunctive cause. This seems a bit ad hoc to me (Garson forthcoming c).

Note that I'm not endorsing the entirety of Dretske's (1986) approach to naturalizing intentionality. There are features I don't accept. He defined a representation's content as whatever the *representation* has the function of carrying information about (p. 22). This is questionable on two counts. First, he defined the content of a representation in terms of the function of that very representation, rather than the function of the representation's producer mechanism. He also thought that for one thing to carry information about another, there must be a very tight correlation between the two things, which I don't accept. I think the best move is to take his strategy for dealing with distal contents and plug it into the broader framework of producer teleosemantics.

There are other problems with teleosemantics that I haven't addressed here, in addition to proximal–distal indeterminacy. Even though I didn't address Burge's problem directly, my view avoids that problem as well. If you remember, Burge (2010), following earlier critics, argues that teleosemantics is incorrect because it attempts to reduce representation, reference and, ultimately, truth to ancestral contributions to fitness. He thinks fitness contributions are just the wrong sort of things for naturalizing representation: After all, false beliefs and other misrepresentations sometimes contribute to fitness, too. As we can see from the above discussion, Burge is relying on an overly limited interpretation of what teleosemantics asserts. Teleosemantics doesn't say that representations are grounded in ancestral contributions to fitness. Rather, representations are grounded, ultimately, in selection processes, but there are other selection processes in the world in addition to natural selection.

As I already said, there are other ways theorists have tried to extend teleosemantics to address the problem of novel contents. These solutions have emphasized the distinction between direct and derived proper functions, the role of compositionality, and the idea that beliefs can undergo something like a function-bestowing selection process. Despite my reservations about the first two, I believe that neural selection can supplement those strategies quite well. In this sense, GSE is one more tool in the teleosemanticists' toolkit for making sense of new representations.

A Programmatic Epilogue

I've said what functions are and why they matter. What are functions? A trait's function is whatever it did in the past that contributed to its differential reproduction, or differential retention, in a population. Why do functions matter? Debates about function play a pivotal role in philosophy of mind, philosophy of medicine, and philosophy of biology. Those debates have a real impact on science, too, such as in biomedicine, psychiatry, and genetics. I've shown that my own view, GSE, pushes those debates further along in specific ways and often with surprising outcomes.

GSE isn't just the latest debut in the burgeoning functions literature; it's also fruitful. It opens a torrent of new questions or helps us pose old questions in a new way. Here, I list (in no particular order) some of the new questions that GSE opens up. These are all matters for future research to answer.

1   I've restricted my attention to functions in biology, and made only occasional forays into psychology – and even there I focused on the most rudimentary psychological operations. I've said next to nothing about functions in the social realm, but they exist there, too: What's the function of judicial punishment? What's the function of capitalism in Western democracy? At the end of Chapter 4 I indicated, following Reid Blackman, that GSE might give us a key to thinking about these issues in a rigorous way. The question is: Are social functions GSE functions, too? Or do we need other theories of function to make sense of them? And if so, what is it about those sorts of functions that resist incorporation into GSE?
2   Along the same lines, how do artifact functions fit in? I've indicated my resistance to incorporating artifact functions into GSE, since, from what I can see, artifacts acquire their functions on their first appearance in the world, long before anything like selection happens (and setting aside the "virtual" selection that happens in the designer's mind). I also

indicated that artifacts could acquire functions through a kind of selection, but, if anything, these would be overlaid on the artifact's first function. Clearly, there's more to be said. How do these two sorts of functions interact to shape the form of artifacts around us?

3   One area where the functions debate looms large is in genetics, particularly regarding the question of what proportion of the genome is "functional." Biologists, largely through the work of geneticists W. Ford Doolittle, Dan Graur, and others, have shown that this is a conceptual problem and not merely an empirical one. How should we think about genome function? Doolittle (2013) has emphasized that there are multiple levels of selection (such as gene selection and clade selection) and that a stretch of DNA could acquire a function by virtue of its participation in any one of these processes. That makes me optimistic that GSE can help us make sense of functions in genetics, perhaps even better than SE does. One reason for that is that GSE would, in theory, allow clade selection to yield new functions despite the fact that clades don't actually reproduce.

4   Connected with genetics, there's a more general issue about thinking through function pluralism. A standard line in philosophy of biology is that there are two senses of "function" in biology, one captured by SE and the other captured by CR; biologists have already started to appropriate this form of pluralism to make sense of their projects (Kellis et al. 2014). I showed that the first sense is better captured by GSE, and I raised questions about whether CR really captures this other sense of function, this "ahistorical" and "minimal" sense. At the very least, the whole issue of pluralism needs to be rethought from the ground up – ideally with the kind of careful sensitivity to practice that Amundson and Lauder (1994) exhibit. Then we'll have a better picture of what the ENCODE debate is actually about.

5   I made a recommendation for how we should go about explaining diseases. I said that, as a rule, we shouldn't seek mechanisms for diseases. Rather, we should explain diseases as the outcomes of broken mechanisms for functions. I made the empirical conjecture that this approach could help us integrate biomedical knowledge effectively, and might even serve as a guide for discovering new things. How plausible is this conjecture? And to what extent does that conjecture actually mirror mainstream biomedical discovery? These are questions that biomedical researchers, philosophers of medicine, and historians of medicine might help answer.

6 I argued that GSE undermines the attempt to characterize mental disorders in terms of biological dysfunctions, since, if GSE is correct, there are probably some mental disorders, or specific symptoms thereof, that aren't dysfunctional. My point, however, wasn't merely negative. I think we can use GSE to help us map out how functions and dysfunctions collide in specific psychiatric contexts. For example, I used GSE to suggest that the delusions of schizophrenia might represent a *functional* response to a low-level perceptual *dysfunction*. I also drew our attention to the prospect that some conditions might represent developmental mismatches, not dysfunctions. Here I glimpse, dimly, the outlines of a research strategy for modeling specific mental disorders.

7 Functions play a pivotal role in the project of naturalizing semantics, and I've shown how GSE can help us solve a specific problem that's dogged teleosemantics – namely, the problem of novel contents. I indicated, fleetingly, that it could potentially solve other objections as well. Some have argued that you can't reduce truth to reproductive success, since false beliefs can be reproductively successful, too. If GSE is right, then representation can't, strictly speaking, be reduced to past reproductive success, and that critique would miss its mark. (As I noted, even a delusional system of beliefs can undergo something like differential retention because of its effects.) It remains a future task to rethink teleosemantics, and some of the criticisms against it, through the lens of GSE, although Garson and Papineau (in prep) are taking initial steps in that direction.

References

Abrams, M. 2005. Teleosemantics without natural selection. *Biology and Philosophy* 20: 97–116.

Adams, F. R. 1979. A goal-state theory of function attributions. *Canadian Journal of Philosophy* 9: 492–518.

Alberts, B., et al. 2012. *Molecular Biology of the Cell*, 5th ed. New York: Garland.

Allen, C., and Bekoff, M. 1995. Biological function, adaptation, and natural design. *Philosophy of Science* 62: 609–622.

American Psychiatric Association. 2013. *Diagnostic and Statistical Manual of Mental Disorders: DSM-5*. Washington DC: American Psychiatric Association.

Amundson, R. 2000. Against normal function. *Studies in History and Philosophy of Biological and Biomedical Sciences* 31:33–53.

Amundson, R., and Lauder, G. V. 1994. Function without purpose: The uses of causal role function in evolutionary biology. *Biology and Philosophy* 9: 443–469.

Antonini, A., and Stryker, M. P. 1993. Development of individual geniculocortical arbors in cat striate cortex and effects of binocular impulse blockade. *Journal of Neuroscience* 13: 3549–3573.

Ayala, F. J. 1970. Teleological explanations in evolutionary biology. *Philosophy of Science* 37: 1–15.

Barlow, H. B. 1988. Neuroscience: A new era? *Nature* 331: 571.

Barnes, E. 2016. *The Minority Body: A Theory of Disability*. Oxford: Oxford University Press.

Barros, B. 2008. Natural selection as a mechanism. *Philosophy of Science* 75:306–322.

Bateson, G., et al. 1956. Toward a theory of schizophrenia. *Behavioral Science* 1: 251–264.

Bechtel, W. and Abrahamsen, A. 2005. Explanation: A mechanist alternative. *Studies in the History and Philosophy of Biological and Biomedical Sciences* 36: 412–441.

Bechtel, W. and Richardson, R. C. 1993. *Discovering Complexity: Decomposition and Localization as Strategies in Scientific Research*. Princeton, NJ: Princeton University Press.

Bedau, M. 1991. Can biological teleology be naturalized? *Journal of Philosophy* 88: 647–655.

1992. Where's the good in teleology? *Philosophy and Phenomenological Research* 52: 781–802.

Bigelow, J. and Pargetter, R. 1987. Functions. *Journal of Philosophy* 84: 181–196.

Bock, W. J. and von Wahlert, G. 1965. Adaptation and the form-function complex. *Evolution* 19: 269–299.

Boorse, C. 1975. On the distinction between disease and illness. *Philosophy and Public Affairs* 5: 49–68.

1976. Wright on functions. *Philosophical Review* 85: 70–86.

1977. Health as a theoretical concept. *Philosophy of Science* 44: 542–573.

2002. A rebuttal on functions. In A. Ariew, R. Cummins, and M. Perlman, eds., *Functions: New Essays in the Philosophy of Psychology and Biology*, Oxford: Oxford University Press, 63–112.

2014. A second rebuttal on health. *Journal of Medicine and Philosophy* 39: 683–724.

Bouchard, F. 2008. Causal processes, fitness, and the differential persistence of lineages. *Philosophy of Science* 75: 560–70.

2013. How ecosystem evolution strengthens the case for function pluralism. In P. Huneman, ed., *Function: Selection and Mechanisms*. Dordrecht: Springer, 83–95.

Bovet, P. and Parnas, J. 1993. Schizophrenic delusions: A phenomenological approach. *Schizophrenia Bulletin* 19: 579–597.

Brandon, R. N. 1990. *Adaptation and Environment*. Princeton, NJ: Princeton University Press.

2013. A general case for functional pluralism. In P. Huneman, ed., *Function: Selection and Mechanisms*. Dordrecht: Springer, 97–104.

Bromberger, S. 1966. Why-questions. In R. Colodny, ed., *Mind and Cosmos*. Pittsburgh: University of Pittsburgh Press., 86–111.

Brown, M. C., Jansen, J. K. S., and Van Essen, D. 1976. Polyneural innervation of skeletal muscle in new-born rats and its elimination during maturation. *Journal of Physiology* 261: 387–422.

Brüne, M. 2008. *Textbook of Evolutionary Psychiatry: The Origins of Psychopathy*. Oxford: Oxford University Press.

Buller, D. J. 1997. Individualism and evolutionary psychology (or, in defense of "narrow" functions). *Philosophy of Science* 64: 74–95.

1998. Etiological theories of function: A geographical survey. *Biology and Philosophy* 13: 505–527.

2002. Function and design revisited. In A. Ariew, R. Cummins, and M. Perlman, eds., *Functions: New Essays in the Philosophy of Psychology and Biology*. Oxford: Oxford University Press, 63–112.

2005. *Adapting Minds: Evolutionary Psychology and the Persistent Quest for Human Nature*. Cambridge, MA: MIT Press.

Burge, T. 2010. *Origins of Objectivity*. Oxford: Oxford University Press.

Burnet, F. M. 1959. *The Clonal Selection Theory of Acquired Immunity*. Cambridge: Cambridge University Press.

Buss, D. M. 2008. *Evolutionary Psychology: The New Science of the Mind*, 3rd ed. Boston: Pearson.

Campbell, D. T. 1960. Blind variation and selective survival as a general strategy in knowledge processes. In M. C. Yovits, and S. Cameron, eds., *Self-Organizing Systems*. New York: Pergamon Press, 205–231.

Carey, N. 2015. *Junk DNA: A Journey through the Dark Matter of the Genome*. Icon: London.

Carleton, R. N., et al. 2013. Intolerance of uncertainty as a contributor to fear and avoidance symptoms of panic attacks. *Cognitive Behaviour Therapy* 42: 328–341

Cao, R. 2012. A teleosemantic approach to information in the brain. *Biology & Philosophy* 27: 49–71.

Caro, T., et al. 2014. The function of zebra stripes. *Nature Communications* 5:3535.

Changeux, J. P. 1985. *Neuronal Man*. New York: Pantheon Books.

Changeux, J. P. and Danchin, A. 1976. Selective stabilization of developing synapses as a mechanism for the specification of neuronal networks. *Nature* 264: 705–711.

Changeux, J. P. and Dehaene, S. 1989. Neuronal models of cognitive functions. *Cognition* 33: 63–109.

Chemero, A. 2009. *Radical Embodied Cognitive Science*. Cambridge, MA: MIT Press.

Chung, W. et al. 2015. Do glia drive synaptic and cognitive impairment in disease? *Nature Neuroscience* 18: 1539–1545.

Clark, D. M. 1986. A cognitive approach to panic disorder. *Behaviour Research and Therapy* 24: 461–470.

1997. Panic disorder and social phobia. In D. M. Clark, and C. G. Fairburn, eds., *Science and Practice of Cognitive Behaviour Therapy*. Oxford: Oxford University Press, 119–153.

Confer, J. C. et al. 2010. Evolutionary psychology: Controversies, questions, prospects, and limitations. *American Psychologist* 65: 110–126

Crane, T. 2003. *The Mechanical Mind*, 2nd ed. London: Routledge.

Craver, C. F. 2001. Role functions, mechanisms, and hierarchy. *Philosophy of Science* 68: 53–74.

2013. Functions and mechanisms: A perspectivalist view. In P. Huneman, ed., *Function: Selection and Mechanisms*. Dordrecht: Springer. 133–158.

Craver, C. F. and Darden, L. 2013. *In Search of Mechanisms: Discoveries across the Life Sciences*. Chicago: University of Chicago Press.

Crick, F. 1989. Neural edelmanism. *Trends in Neurosciences* 12: 240–48.

Crowley, J. C. and Katz, L. C. 1999. Development of ocular dominance columns in the absence of retinal input. *Nature Neuroscience* 2: 1125–1130.

2000. Early development of ocular dominance columns. *Science* 290: 1321–1324.

Cummins, R. 1975. Functional analysis. *Journal of Philosophy* 72: 741–765.

1983. *The Nature of Psychological Explanation*. Cambridge, MA: MIT Press.

Cziko, G. 1995. *Without Miracles: Universal Selection Theory and the Second Darwinian Revolution*. Cambridge, MA: MIT Press.

Darden, L. 2006. *Reasoning in Biological Discoveries*. Cambridge: Cambridge University Press.

Darden, L. and Cain, J. A. 1989. Selection type theories. *Philosophy of Science* 56: 106–129.

Darwin, C. 1871. *The Descent of Man, and Selection in Relation to Sex. Volume 2*. London: John Murray.

Davies, P.S. 2001. *Norms of Nature: Naturalism and the Nature of Functions*. Cambridge, MA: MIT Press.

Dawkins, R. 1982. *The Extended Phenotype: The Gene as the Unit of Selection*. Oxford: Oxford University Press.

Deacon, T. 1997. *The Symbolic Species*. New York: W. W. Norton.

Dennett, D. C. 1987. *The Intentional Stance*. Cambridge, MA: MIT Press.

1996. *Kinds of Minds*. New York: Basic Books.

Deppmann, D. et al. 2008. A model of neuronal competition during development. *Science* 320: 369–373.

Dickinson, A. and Balleine, B. 1994. Motivational control of goal-directed action. *Animal Learning and Behavior* 22: 1–18.

Doolittle, W. F. 2013. Is junk DNA bunk? A critique of ENCODE. *Proceedings of the National Academy of Sciences of the United States of America* 110: 5294–5300.

2014. Natural selection through survival alone, or the *possibility* of Gaia. *Biology and Philosophy* 29: 415–423.

2017. Making the most of clade selection. *Philosophy of Science* 84: 275–295.

Dover, G. 2000. *Dear Mr. Darwin: Letters on the Evolution of Life and Human Nature*. Berkeley: University of California Press.

Dretske, F. 1973. Contrastive statements. *Philosophical Review* 81: 411–437.

1981. *Knowledge and the Flow of Information*. Cambridge, MA: MIT Press.

1986. Misrepresentation. In R. Bogdan, ed., *Belief: Form, Content, and Function*. Oxford: Clarendon Press, 17–36.

Eddy, S. R. 2012. The C-value paradox, junk DNA and ENCODE. *Current Biology* 22: R898–R899.

Edelman, G. M. 1987. *Neural Darwinism: The Theory of Neuronal Group Selection*. New York: Basic Books.

Eldakar, O. T., et al. 2010. The role of multilevel selection in the evolution of sexual conflict in the water strider *Aquarius remigis*. *Evolution* 64: 3183–3189.

Elliott, T and Shadbolt, N. R. 1998. Competition for neurotrophic factors: Ocular dominance columns. *Journal of Neuroscience* 18: 5850–5858.

Enç, B. 2002. Indeterminacy of function attributions. In. A. Ariew, R. Cummins, and M. Perlman, eds., *Functions: New Essays in the Philosophy of Psychology and Biology*. Oxford: Oxford University Press, 291–313.

ENCODE Project Consortium. 2012. An integrated encyclopedia of DNA elements in the human genome. *Nature* 489: 57–74.

Faucher, L. and Blanchette, I. 2011. Fearing new dangers: Phobias and the cognitive complexity of human emotions. In R. Adriaens, and A. De Block. eds., *Maladapting Minds: Philosophy, Psychiatry, and Evolutionary Theory*. Oxford: Oxford University Press, 33–64.

Feinberg, I. 1982/83. Schizophrenia: Caused by a fault in programmed synaptic elimination during adolescence? *Journal of Psychiatric Research* 17: 319–334.

First, M. B. 2007. Potential implications of the harmful dysfunction analysis for the development of DSM-V and ICD 11. *World Psychiatry* 6: 158–159.

Fodor, J. A. 1984. Semantics, Wisconsin style. *Synthese* 59: 231–250.

1987. *Psychosemantics*. Cambridge, MA: MIT Press.

1990. *A Theory of Content and Other Essays*. Cambridge, MA: MIT Press.

Foucault, M. 2006 [1961]. *History of Madness*. London: Routledge.

Frankland, S. M. and Greene, J. D. 2015. An architecture for encoding sentence meaning in left mid-superior temporal cortex. *Proceedings of the National Academy of Sciences* 112: 11732–11737.

Fromm-Reichmann, F. 1948. Notes on the development of treatment of schizophrenics by psychoanalytic psychotherapy. *Psychiatry* 11: 263–273.

Garson, J. 2006. *Psychiatric Disorders and Biological Dysfunctions: Some Philosophical Questions Concerning Psychiatry*. Dissertation, University of Texas at Austin.

2010. Schizophrenia and the dysfunctional brain. *Journal of Cognitive Science* 11: 215–246.

2011. Selected effects functions and causal role functions in the brain: The case for an etiological approach to neuroscience. *Biology and Philosophy* 26: 547–565.

2012. Function, selection, and construction in the brain. *Synthese* 189: 451–481.

2013. The functional sense of mechanism. *Philosophy of Science* 80: 317–333.

2015. *The Biological Mind: A Philosophical Introduction*. London: Routledge.

2016. *A Critical Overview of Biological Functions*. Dordrecht: Springer.

2017a. A generalized selected effects theory of function. *Philosophy of Science* 84: 523–543.

2017b. Against organizational functions. *Philosophy of Science* 84: 1093–1103.

2017c. A "model schizophrenia": Amphetamine psychosis and the transformation of American psychiatry. In S. Casper, and D. Gavrus, eds., *The History of the Brain and Mind Sciences: Technique, Technology, Therapy*. Rochester, NY: University of Rochester Press, 202–228.

2017d. Mechanisms, Phenomena, and Functions. In S. Glennan and P. Illari, eds., *Routledge Handbook of Philosophy of Mechanisms*. London: Routledge, 104–115.

Forthcoming a. How to be a function pluralist. *British Journal for the Philosophy of Science*.

Forthcoming b. The developmental plasticity challenge to Wakefield's view. In L. Faucher and D. Forest, eds., *Defining Mental Disorder: Jerome Wakefield and His Critics*. Cambridge, MA: MIT Press.

Forthcoming c. Do constancy mechanisms save distal content? *Philosophical Quarterly*.

Garson, J. and Papineau, D. In prep. Teleosemantics, selection, and learning.

Garson, J. and Piccinini, G. 2014. Functions must be performed at appropriate rates in appropriate situations. *British Journal for the Philosophy of Science* 65:1–20.

Gazzaniga, M. S. 1992. *Nature's Mind: The Biological Roots of Thinking, Emotions, Sexuality, Language, and Intelligence*. New York: Basic Books.

Gibson, D. A. and Ma, L. 2011. Developmental regulation of axon branching in the vertebrate nervous system. *Development* 138: 183–195.

Glennan, S. 1996. Mechanisms and the nature of causation. *Erkenntnis* 44:49–71.

2005. Modeling mechanisms. *Studies in History and Philosophy of Biological and Biomedical Sciences* 36: 443–464.

2017. *The New Mechanical Philosophy*. Oxford: Oxford University Press.

Glover, V. 2011. Prenatal stress and the origins of psychopathology: An evolutionary perspective. *Journal of Child Psychology and Psychiatry* 52: 356–367.

Gluckman, P. and Hanson, M. eds. 2006. *Developmental Origins of Health and Disease*. Oxford: Oxford University Press.

Godfrey-Smith, P. 1989. Misinformation. *Canadian Journal of Philosophy* 19: 533–550.

1992. Indication and adaptation. *Synthese* 92: 283–312.

1993. Functions: Consensus without unity. *Pacific Philosophical Quarterly* 74: 196–208.

1994. A modern history theory of functions. *Nous* 28: 344–362.

2000. Information, arbitrariness, and selection: Comments on Maynard Smith. *Philosophy of Science* 67: 202–207.

2009. *Darwinian Populations and Natural Selection*. Oxford: Oxford University Press.

Goode, R. and Griffiths, P. E. 1995. The misuse of Sober's selection of/selection for distinction. *Biology and Philosophy* 10: 99–108.

Gould, S. J., and Lewontin, R. 1979. The Spandrels of San Marco and the panglossian paradigm. *Proceedings of the Royal Society of London* 205: 281–288.

Grace, A. A. 2000. Gating of information flow within the limbic system and the pathophysiology of schizophrenia. *Brain Research Reviews* 31: 330–341.

Graham, G. 2010. *The Disordered Mind*. London: Routledge.

Grantham, T. 2001. Do operant behaviors replicate? *Behavioral and Brain Sciences* 24: 538–539.

Graur, D. et al. 2013. On the immortality of television sets: "Function" in the human genome according to the evolution-free gospel of ENCODE. *Genome Biology and Evolution* 5:578–590.

Griffiths, P. E. 1992. Adaptive explanation and the concept of a vestige. In P. Griffiths, ed., *Trees of Life: Essays in Philosophy of Biology*. Dordrecht: Kluwer, 111–131.

1993. Functional analysis and proper function. *British Journal for the Philosophy of Science* 44: 409–422.

2006. Function, homology, and character individuation. *Philosophy of Science* 73: 1–25.

Hardcastle, V. G. 1999. Understanding functions: A pragmatic approach. In V. G. Hardcastle, ed., *Where Biology Meets Psychology: Philosophical Essays.* Cambridge, MA: MIT Press, 27–43.

Harper, K. L., et al. 2016. Mechanism of early dissemination and metastasis in Her2$^+$ mammary cancer. *Nature* 540: 588–592.

Harris, R. H. T. P. 1930. *Report on the Bionomics of the Tsetse Fly (Glossina pallidipes Aust.) and a Preliminary Report on a New Method of Control.* Peitermaritzburg: Provincial Administration of Natal.

Hausman, D. 2011. Is an overdose of paracetamol bad for one's health? *British Journal for the Philosophy of Science* 62: 657–668.

Havstad, J. C. 2011. Problems for natural selection as a mechanism. *Philosophy of Science* 78:512–523.

Hebb, D. O. 1949. *The Organization of Behavior.* New York: Wiley.

Heim, C, and Nemeroff, C. B. 2001. The role of childhood trauma in the neurobiology of mood and anxiety disorders: Preclinical and clinical studies. *Biological Psychiatry* 49: 1023–1039.

Helgeson, C. 2015. There is no asymmetry of identity assumptions in the debate over selection and individuals. *Philosophy of Science* 82: 21–31.

Hempel, C. G. 1965. *Aspects of Scientific Explanation.* New York: Free Press.

Higham, T. E. et al. 2017. Rattlesnakes are extremely fast and variable when striking at kangaroo rats in nature: Three-dimensional high-speed kinematics at night. *Scientific Reports* 7: 40412.

Himsworth, H. P. 1939. The mechanism of diabetes mellitus. *The Lancet* 234: 171–176.

Horn, D. and McCulloch, R. 2010. Molecular mechanisms underlying the control of antigenic variation in African trypanosomes. *Current Opinion in Microbiology* 13: 700–705.

Hruby, G. G. and Goswami, U. 2011. Neuroscience and reading: A review for reading education researchers. *Reading Research Quarterly* 46: 156–172.

Hull, D. L., Langman, R. E., and Glenn, S. S. 2001. A general account of selection: Biology, immunology and behavior. *Behavioral and Brain Sciences* 24: 511–527.

Huttenlocher, P. R. 1979. Synaptic density in the human frontal cortex: Developmental changes and effects of aging. *Brain Research* 163: 195–205.

Innocenti, G. M. and Price, D. J. 2005. Exuberance in the development of cortical networks. *Nature Reviews Neuroscience* 6: 955–965.

Isbell, L. A. 2009. *The Fruit, the Tree, and the Serpent: Why We See so Well.* Cambridge, MA: Harvard University Press.

Jablonka, E. and Lamb, M. J. 2005. *Evolution in Four Dimensions.* Cambridge, MA: MIT Press.

Jacob, P. 1997. *What Minds Can Do: Intentionality in a Non-Intentional World.* Cambridge: Cambridge University Press.

Jerne, N. 1955. The natural-selection theory of antibody formation. *Proceedings of the National Academy of Sciences of the United States of America* 41: 849–857.

Kandel, E. R., et al. 2013. *Principles of Neural Science,* 5th ed. New York: McGraw Hill.

Katz, L.C. and Shatz, C. J. 1996. Synaptic activity and the construction of cortical circuits. *Science* 234: 1133–1138.

Kauer, J. A. and Malenka, R. C.. 2007. Synaptic plasticity and addiction. *Nature Reviews Neuroscience* 8:844–858.

Kelemen, D. 1999. The scope of teleological thinking in preschool children. *Cognition* 70: 241–272.

Kellis, M., et al. 2014. Reply to Brunet and Doolittle: Both selected effect and causal role elements can influence human biology and disease. *Proceedings of the National Academy of Sciences of the United States of America.* 111: E3366.

Kingma, E. 2010. Paracetamol, poison, and polio: Why Boorse's account of function fails to distinguish health and disease. *British Journal for the Philosophy of Science* 61: 241–264.

2015. Situation-specific disease and dispositional function. *British Journal for the Philosophy of Science* 67: 391–404.

Kingsbury, J. 2006. A proper understanding of Millikan. *Acta Analytica* 21: 23–40.

2008. Learning and selection. *Biology and Philosophy* 23: 493–507.

Kirk, S. A and Kutchins, H. 1992. *The Selling of DSM.* New York: Aldine de Gruyter.

Klein, D. F. 1978. A proposed definition of mental illness. In. R. L. Spitzer, and D. F. Klein, eds., *Critical Issues in Psychiatric Diagnosis.* New York: Raven Press, 47–71.

1999. Harmful dysfunction, disease, illness, and evolution. *Journal of Abnormal Psychology* 108: 421–429.

Kraemer, D. M. 2013. Statistical theories of functions and the problem of epidemic disease. *Biology and Philosophy* 28: 423–438.

2014. Revisiting recent etiological theories of functions. *Biology and Philosophy* 29: 747–759.

Krenn, H. and Aspöck, H. 2012. Form, function, and evolution of the mouthparts of blood-feeding Arthropoda. *Arthropod Structure and Development* 41: 101–118.

Kriegel, U. 2013. The phenomenological intentionality research program. In U. Kriegel, ed., *Phenomenological Intentionality.* Oxford: Oxford University Press. 1–26.

Kripke, S. 1980. *Naming and Necessity.* Cambridge, MA: Harvard University Press.

Krohs, U. and Kroes, P. eds. 2009. *Functions in Biological and Artificial Worlds.* Cambridge, MA: MIT Press.

Lakatos, I. 1970. Falsification and the methodology of scientific research programmes. In I. Lakatos, and A. Musgrave, eds., *Criticism and the Growth of Knowledge*. Cambridge: Cambridge University Press. 91–195.

Landsberg, J., et al. 2012. Melanomas resist T-cell therapy through inflammation-induced reversible de-differentiation. *Nature* 490: 412–416.

Larison, B., et al. 2015. How the zebra got its stripes: A problem with too many solutions. *Royal Society Open Science*. DOI: 10.1098/rsos.140452.

Lettvin, J. Y., et al. 1959. What the frog's eye tells the frog's brain. *Proceedings of the IRE* 47: 1940–1951.

LeVay, S., Stryker, M. P., and Shatz, C. J. 1978. Ocular dominance columns and their development in layer IV of the cat's visual cortex: A quantitative study. *Journal of Comparative Neurology* 179: 223–244.

LeVay, S., Wiesel, T. N., and Hubel, D. H. 1980. The development of ocular dominance columns in normal and visually deprived monkeys. *Journal of Comparative Neurology* 191: 1–51.

Levy, A. 2013. Three kinds of new mechanism. *Biology and Philosophy* 28: 99–114.

Lewens, T. 2001. Sex and selection: A reply to Matthen. *British Journal for the Philosophy of Science* 52: 589–598.

2004. *Organisms and Artifacts: Design in Nature and Elsewhere*. Cambridge, MA: MIT Press.

Lewis, D. K. 1973. *Counterfactuals*. Oxford: Basil Blackwell.

1986. *On the Plurality of Worlds*. Cambridge, MA: Harvard University Press/ Oxford: Basil Blackwell.

Lewontin, R. C. 1998. The evolution of cognition: Questions we will never answer. In D. Scarborough, and S. Sternberg, eds., *An Invitation to Cognitive Science, vol 4: Methods, Models, and Conceptual Issues*, 2nd ed. Cambridge, MA: MIT Press, 107–132.

Lichtman, J. W., Burden, S. J., Culican, S. M., and Wong, R. O. L. 1999. Synapse formation and elimination. In M. J. Zigmond, F. E. Bloom, S. C. Landis, J. L. Roberts, and L. R. Squire, eds., *Fundamental neuroscience*. San Diego: Academic Press, 547–580.

Lilienfeld, S. O., and Marino, L. 1995. Mental disorder as a Roschian concept: A Critique of Wakefield's "harmful dysfunction" analysis. *Journal of Abnormal Psychology* 104: 411–420.

1999. Essentialism revisited: Evolutionary theory and the concept of mental disorder. *Journal of Abnormal Psychology* 108: 400–411.

Loewer, B., 1987. From information to intentionality. *Synthese* 70: 287–317.

Lorenz, K. 2002 [1963]. *On Aggression*. New York: Routledge.

Ludewig, S. et al. 2005. Information-processing deficits and cognitive dysfunction in panic disorder. *Journal of Psychiatry and Neuroscience* 30: 37–43.

Macdonald, D. W. 2009. *Encyclopedia of Mammals*. Oxford: Oxford University Press.

Machamer, P., Darden, L., and Craver, C. F. 2000. Thinking about mechanisms. *Philosophy of Science* 67:1–25.

Maclaurin, J. and Sterelny, K. *What Is Biodiversity?* Chicago: University of Chicago Press.

Maley, C. J. and Piccinini, G. 2018. A unified mechanistic account of teleological functions for psychology and neuroscience. In D. Kaplan, ed., *Integrating Psychology and Neuroscience: Prospects and Problems*. Oxford: Oxford University Press, 236–256.

Marten, G. G. 2001. *Human Ecology: Basic Concepts for Sustainable Development*. London: Earthscan.

Matthen, M. 1999. Evolution, Wisconsin style: Selection and the explanation of individual traits. *British Journal for the Philosophy of Science* 50: 143–150.

2002. Origins are not essences in evolutionary systematics. *Canadian Journal of Philosophy* 32: 167–181.

Matthen, M. and Ariew, A. 2002. Two ways of thinking about fitness and natural selection. *Journal of Philosophy* 99: 55–83.

Matthews, L. 2016. On closing the gap between philosophical concepts and their usage in scientific practice: A lesson from the debate about natural selection as a mechanism. *Studies in History and Philosophy of Biological and Biomedical Sciences* 55: 21–28.

Matthewson, J. 2015. Defining paradigm darwinian populations. *Philosophy of Science* 82: 178–197.

Maynard Smith, J. 1990. Explanation in biology. In D. Knowles, ed., *Explanation and its Limits*. Cambridge: Cambridge University Press, 65–72.

2000. The concept of information in biology. *Philosophy of Science* 67: 177–194.

McGowan, P. O. et. al. 2009. Epigenetic regulation of the glucocorticoid receptor in human brain associates with childhood abuse. *Nature Neuroscience* 12: 342–348.

McLaughlin, P. 2001. *What Functions Explain: Functional Explanation and Self-Reproducing Systems*. Cambridge: Cambridge University Press.

Meyer, R. L., and Sperry, R. W. 1976. Retinotectal specificity: Chemoaffinity theory. In G. Gottlieb, ed., *Studies on the Development of Behavior and the Nervous System. Vol. 3: Neural and Behavioral Specificity*. New York: Academic Press, 111–149.

Mill, J. S. 1882. *A System of Logic*, 8th ed. New York: Harper and Brothers.

Millikan, R. G. 1984. *Language, Thought, and Other Biological Categories*. Cambridge, MA: MIT Press.

1989a. In defense of proper functions. *Philosophy of Science* 56: 288–302.

1989b. Biosemantics. *Journal of Philosophy* 86: 281–297.

1993. *White Queen Psychology and Other Essays for Alice*. Cambridge, MA: MIT Press.

2004. *Varieties of Meaning*. Cambridge, MA: MIT Press.

2013. Reply to Neander. In D. Ryder, J. Kingsbury, and K. Williford, eds., *Millikan and Her Critics*. Malden, MA: Wiley-Blackwell, 37–40.

Millstein, R. L. 2009. Populations as individuals. *Biological Theory* 4: 267–273.

Moghaddam-Taaheri, S. 2011. Understanding pathology in the context of physiological mechanisms: The practicality of a broken-normal view. *Biology and Philosophy* 26: 603–611.

Moreno, A. and Mossio, M. 2015. *Biological Autonomy: A Philosophical and Theoretical Inquiry.* Dordrecht: Springer.

Moss, L. 2012. Is the philosophy of mechanism philosophy enough? *Studies in the History and Philosophy of Biological and Biomedical Sciences* 43: 164–172.

Mossio, M., Saborido, C., and Moreno, A. 2009. An organizational account for biological functions. *British Journal for the Philosophy of Science* 60: 813–841.

Murphy, D. 2005. Can evolution explain insanity? *Biology & Philosophy* 20: 745–766.

Murphy, D., and S. Stich. 2000. Darwin in the madhouse: Evolutionary psychology and the classification of mental disorders. In P. Carruthers and A. Chamberlain, eds., *Evolution and the Human Mind: Modularity, Language, and Meta-Cognition.* Cambridge: Cambridge University Press, 62–92.

Nagel, E. 1953. Teleological explanation and teleological systems. In S. Ratner, ed., *Vision and Action.* New Brunswick, NJ: Rutgers University Press, 537–558.

1961. *The Structure of Science,* New York: Harcourt, Brace and World.

Nanay, B. 2013. Success semantics: The sequel. *Philosophical Studies* 165: 151–165.

Neander, K. 1983. *Abnormal Psychobiology.* Dissertation, La Trobe University, Bundoora, Australia.

1988. What does natural selection explain? Correction to Sober. *Philosophy of Science* 55: 422–426.

1991. Functions as selected effects: The conceptual analyst's defense. *Philosophy of Science* 58: 168–184.

1995a. Pruning the tree of life. *British Journal for the Philosophy of Science* 46: 59–80.

1995b. Misrepresenting and malfunctioning. *Philosophical Studies* 79: 109–141.

1999. Fitness and the fate of unicorns. In. V. G. Hardcastle, ed., *Where Biology Meets Psychology: Philosophical Essays.* Cambridge, MA: MIT Press, 3–26

2012. Teleosemantic theories of mental content. *Stanford Encyclopedia of Philosophy.* http://plato.stanford.edu/entries/content-teleological/

2013. Toward an informational teleosemantics. In D. Ryder, J. Kingsbury, and K. Williford, eds., *Millikan and her Critics.* Malden, MA: Wiley-Blackwell, 21–36.

2017a. Functional analysis and the species design. *Synthese* 194: 1147–1168.

2017b. *A Mark of the Mental: In Defense of Informational Teleosemantics.* Cambridge, MA: MIT Press.

Neander, K., and Rosenberg, A. 2012. Solving the circularity problem for functions. *Journal of Philosophy* 109: 613–622.

Nesse, R. M. 2000. Is depression an adaptation? *Archives of General Psychiatry* 57: 14–20.

2007. Evolution is the scientific foundation for diagnosis: Psychiatry should use it. *World Psychiatry* 6: 160–161.

Okasha, S. 2006. *Evolution and the Levels of Selection*. Oxford: Oxford University Press.

Papineau, D. 1984. Representation and explanation. *Philosophy of Science* 51: 550–572.

1987. *Reality and Representation*. Oxford: Blackwell.

1993. *Philosophical Naturalism*. Oxford: Blackwell.

1997. Teleosemantics and indeterminacy. *Australasian Journal of Philosophy* 76: 1–14.

2006. Naturalist theories of meaning. In E. Lepore and B. C. Smith, eds., *The Oxford Handbook of Philosophy of Language*. Oxford: Oxford University Press, 175–188.

2017. Teleosemantics. In D. L. Smith, ed., *How Biology Shapes Philosophy*. Cambridge: Cambridge University Press, 95–120.

Peacocke, C. 1992. *A Study of Concepts*. Cambridge, MA: MIT Press.

Piccinini, G. 2010. The mind as neural software? Understanding functionalism, computationalism, and functional computationalism. *Philosophy and Phenomenological Research* 81:269–311.

2015. *Physical Computation: A Mechanistic Account*. Oxford: Oxford University Press.

Piccinini, G., and Craver, C. 2011. Integrating psychology and neuroscience: Functional analyses as mechanism sketches. *Synthese* 183: 283–311.

Pietroski, P. M. 1992. Intentionality and teleological error. *Pacific Philosophical Quarterly* 73: 267–281.

Pigliucci, M. 2001. *Phenotypic Plasticity: Beyond Nature and Nurture*. Baltimore, MD: Johns Hopkins University Press.

Pittendrigh, C. S. 1958. Adaptation, natural selection, and behavior. In A. Roe and G. G. Simpson, eds., *Behavior and Evolution*. New Haven: Yale University Press, 390–416.

Plantinga, A. 1993. *Warrant and Proper Function*. Oxford: Oxford University Press.

Preston, B. 1998. Why is a wing like a spoon? A pluralist theory of function. *Journal of Philosophy* 95: 215–254.

Price, C. 2001. *Functions in Mind: A Theory of Intentional Content*. Oxford: Clarendon Press.

Price, D. J., Jarman, A. P., Mason, J. O., and Kind, P. C. 2011. *Building Brains: An Introduction to Neural Development*. Chichester: Wiley-Blackwell.

Purves, D. 1988. A new theory of brain function. *Quarterly Review of Biology* 63: 202–204.

1994. *Neural Activity and the Growth of the Brain*. Cambridge: Cambridge University Press.

Purves, D. and Lichtman, J. W. 1980. Elimination of synapses in the developing nervous system. *Science* 210: 153–157.

Purves, D., White, L. E., and Riddle, D. R. 1996. Is neural development Darwinian? *Trends in Neuroscience* 19: 460–464.

Pust, J. 2001. Natural selection explanation and origin essentialism. *Canadian Journal of Philosophy* 31: 201–220.

Quartz, S. R. 1999. The constructivist brain. *Trends in Cognitive Sciences* 3: 48–57.

Quartz, S. R. and Sejnowski, T. J. 1997. The neural basis of cognitive development: A constructivist manifesto. *Behavioral and Brain Sciences* 20: 537–596.

Rakic, P. 1976. Prenatal genesis of connections subserving ocular dominance in the rhesus monkey. *Nature* 261: 467–471.

Ramsey, W. M. 2007. *Representation Reconsidered*. Cambridge: Cambridge University Press.

Rauschecker, J. P. 1995. Compensatory plasticity and sensory substitution in the cerebral cortex. *Trends in Neurosciences* 18: 36–43.

Richardson, R. C. 2007. *Evolutionary Psychology as Maladapted Psychology*. Cambridge, MA: MIT Press.

Richters, J. and Hinshaw, S. 1999. The abduction of disorder in psychiatry. *Journal of Abnormal Psychiatry* 108: 438–445.

Ridley, M. 2015. *The Evolution of Everything: How New Ideas Emerge*. New York: HarperCollins.

Rogers, D. S. and Ehrlich, P. R. 2008. Natural selection and cultural rates of change. *Proceedings of the National Academy of Sciences of the United States of America*. 105: 3416–3420.

Rogers, P. C. and St. Clair, S. B. 2016. Quaking aspen in Utah: Integrating recent science with management. *Rangelands* 38: 266–272.

Rolli, C. G., et al. 2010. Impact of tumor cell cytoskeleton organization on invasiveness and migration: A microchannel-based approach. *PLoS ONE* 5: e8726.

Rosenberg, A. 2018. Making mechanism interesting. *Synthese* 95:11–33.

Rosenberg, A and Neander, K. 2009. Are homologies (selected effect or causal role) function free? *Philosophy of Science* 76: 307–334.

Rowan, M. J., Klyubin, I., Cullen, W. K., and Anwyl, R. 2003. Synaptic plasticity in animal models of early alzheimer's disease. *Philosophical Transactions of the Royal Society of London B* 358:821–828.

Ruse, M. E. 1973. *The Philosophy of Biology*. Atlantic Highlands, NJ: Humanities Press.

2002. Evolutionary biology and teleological thinking. In A. Ariew, R. Cummins, and M. Perlman, eds., *Functions: New Essays in the Philosophy of Psychology and Biology*. Oxford: Oxford University Press, 33–62.

Saborido, C. and Moreno, A. 2015. Biological pathology from an organizational perspective. *Theoretical Medicine and Bioethics* 36: 83–95.

Saborido, C., Mossio, M., and Moreno, A. 2011. Biological organization and cross-generation functions. *British Journal for the Philosophy of Science* 62: 583–606.

Salkovskis, P. M. 1991. The importance of behaviour in the maintenance of anxiety and panic: A cognitive account. *Behavioural and Cognitive Psychotherapy* 19: 6–19.

Salmon, W. 1989. *Four Decades of Scientific Explanation*. Pittsburgh: University of Pittsburgh Press.

Sarkar, S. 1998. *Genetics and Reductionism*. Cambridge: Cambridge University Press.

2013. Information in animal communication: When and why does it matter? In U. Stegmann, ed., *Animal Communication Theory: Information and Influence*. Cambridge: Cambridge University Press, 189–205.

Schaffner, K. 1993. *Discovery and Explanation in Biology and Medicine*. Chicago: University of Chicago Press.

Schlosser, G. 1998. Self-re-production and functionality: A systems-theoretical approach to teleological explanation. *Synthese* 116: 303–354.

Schulte, P. 2012. How frogs see the world: Putting Millikan's teleosemantics to the test. *Philosophia* 40:483–496.

2018. Perceiving the world outside: How to solve the distality problem for informational teleosemantics. *The Philosophical Quarterly* 68: 349–369.

Schulz, A. 2018. *Efficient Cognition: The Evolution of Representational Decision-Making*. Cambridge, MA: MIT Press.

Schultz, W. and Dickinson, A. 2000. Neuronal coding of prediction errors. *Annual Review of Neuroscience* 23: 473–500.

Schwartz, P. H. 1999. Proper function and recent selection. *Philosophy of Science* 66: S210–S222.

2002. The continuing usefulness account of proper function. In A. Ariew, R. Cummins, and M. Perlman, eds., *Functions: New Essays in the Philosophy of Psychology and Biology*. Oxford: Oxford University Press, 244–260.

2007. Defining dysfunction: Natural selection, design, and drawing a line. *Philosophy of Science* 74: 364–385.

2014. Reframing the disease debate and defending the biostatistical theory. *Journal of Medicine and Philosophy* 39: 572–589.

Sedgwick, P. 1981. Illness: Mental and otherwise. In A. L. Caplan, H. T. Engelhardt, Jr., and J. J. McCartney, eds., *Concepts of Health and Disease: Interdisciplinary Perspectives*. London: Addison-Wesley, 119–29.

Seyfarth, R. M., Cheney, D. L., and Marler, P. 1980. Vervet monkey alarm calls: Semantic communication in a free-ranging primate. *Animal Behaviour* 28: 1070–1094.

Shea, N. 2007. Consumers need information: Supplementing Teleosemantics with an Input Condition. *Philosophy and Phenomenological Research* 75: 404–435.

2013. Inherited representations are read in development. *British Journal for the Philosophy of Science* 64: 1–31.

Skipper, R. A. and Millstein, R. L. 2005. Thinking about evolutionary mechanisms: Natural selection. *Studies in the History and Philosophy of Biological and Biomedical Sciences* 36:327–347.

Smaldino, P. E. and McElreath, R. 2016. The natural selection of bad science. *Royal Society Open Science* 3: 160384.

Snyder, S. H. 1973. Amphetamine psychosis: A "model" schizophrenia mediated by catecholamines. *American Journal of Psychiatry* 130: 61–67.

Sober, E. 1984. *The Nature of Selection*. Chicago: University of Chicago Press.

1995. Natural selection and distributive explanation: A reply to Neander. *British Journal for the Philosophy of Science* 46: 384–397.

Sober, E. and Wilson, D. S. 1998. *Unto Others: The Evolution and Psychology of Unselfish Behavior*. Cambridge, MA: Harvard University Press.

Spector F., and Maurer, M. 2009. Synesthesia: A new approach to understanding the development of perception. *Developmental Psychology* 45: 175–189.

Sperry, R. W. 1965. Embryogenesis of behavioral nerve nets. In R. L. Dehaan and H. Urspring, eds., *Organogenesis*. New York: Holt, Rinehart, and Winston, 161–185.

Spitzer, R. L. 1999. Harmful dysfunction and the DSM analysis of mental disorder. *Journal of Abnormal Psychology* 108: 430–432.

Spitzer, R. L. and Endicott, J. 1978. Medical and mental disorder: Proposed definition and criteria. In R. L. Spitzer, and D. F. Klein, eds., *Critical Issues in Psychiatric Diagnosis*. New York: Raven Press,15–39.

Stegmann, U. 2009. A consumer-based teleosemantics for animal signals. *Philosophy of Science* 76: 864–875.

Stephan, A. H., Barres, B. A., and Stevens, B. 2012. The complement system: An unexpected role in synaptic pruning during development and disease. *Annual Review of Neuroscience* 35: 369–389.

Sterelny, K. 1990. *The Representational Theory of Mind*. Oxford: Blackwell.

1995. Basic minds. *Philosophical Perspectives* 9: 251–270.

2000. The "genetic program" program: A commentary on Maynard Smith on information in biology. *Philosophy of Science* 67: 195–201.

2003. *Thought in a Hostile World: The Evolution of Human Cognition*. Oxford: Blackwell.

Strawson, G. 2008. Real intentionality 3: Why intentionality entails consciousness. In G. Strawson, ed., *Real Materialism and Other Essays*. Oxford: Oxford University Press, 281–305.

Tooby, J. and Cosmides, L. 2006. Toward mapping the evolved functional organization of mind and brain. In E. Sober, ed., *Conceptual Issues in Evolutionary Biology*, 3rd ed.Cambridge, MA: MIT Press, 175–95.

Tremblay, M., Lowery, R. L., and Majewska, A K. 2010. Microglial interactions with synapses are modulated by visual experience. *PIOS Biology* 8: e1000527.

Uhlhaas, P. J. and Mishara, A. L. 2007. Perceptual anomalies in schizophrenia: Integrating phenomenology and cognitive neuroscience. *Schizophrenia Bulletin* 33: 142–156.

Varela, F. 1979. *Principles of Biological Autonomy*. Elsevier: North Holland.

Wagner, A. D., Bunge, S. A., and Badre, D. 2004. Cognitive control, semantic memory, and priming: Contributions from prefrontal cortex. In M.

Gazzaniga, ed., *The Cognitive Neurosciences*, 3rd ed. Cambridge, MA: MIT Press. 709–726.

Wakefield, J. C. 1992. The concept of mental disorder: On the boundary between biological facts and social values. *American Psychologist* 47: 373–388.

1999a. Evolutionary versus prototype analyses of the concept of disorder. *Journal of Abnormal Psychology* 108: 374–399.

1999b. Mental disorder as a black box essentialist concept. *Journal of Abnormal Psychology* 108: 465–472.

2000. Spandrels, vestigial organs, and such: Reply to Murphy and Woolfolk's "The harmful dysfunction analysis of mental disorder." *Philosophy, Psychiatry, and Psychology* 7: 253–269.

Walsh, D. M. 1996. Fitness and function. *British Journal for the Philosophy of Science* 47: 553–574.

1998. The scope of selection: Sober and Neander on what natural selection explains. *Australasian Journal of Philosophy* 76: 250–264.

Walsh, D. M. and Ariew, A. 1996. A taxonomy of functions. *Canadian Journal of Philosophy* 26: 493–514.

Wells, J. 2011. *The Myth of Junk DNA*. Seattle: Discovery Institute Press.

Wiesel, T. N. and Hubel, D. H. 1963. Single-cell responses in striate cortex of kittens deprived of vision in one eye. *Journal of Neurophysiology* 26: 1003–1017.

Weisel, T. N. and Hubel, D. H. 1965. Comparison of the effects of unilateral and bilateral eye closure on cortical unit responses in kittens. *Journal of Neurophysiology* 28: 1029–1040.

Wiggs, C. L. and Martin, A. 1998. Properties and mechanisms of perceptual priming. *Current Opinion in Neurobiology* 8: 227–233.

Williams, G. C. 1966. *Adaptation and Natural Selection: A Critique of Some Current Evolutionary Thought*. Princeton: Princeton University Press.

Wilson, D. S. 2002. *Darwin's Cathedral: Evolution, Religion, and the Nature of Society*. Chicago: University of Chicago Press.

Wimsatt, W. C. 1972. Teleology and the logical structure of function statements. *Studies in the History and Philosophy of Science* 3: 1–80.

Witten, I. B., et al. 2011. Recombinase-driver rat lines: tools, techniques, and optogenetic application to dopamine-mediated reinforcement. *Neuron* 72: 721–733.

Wong R. O. L. and Lichtman J. W. 2002. Synapse elimination. In. L. R. Squire, F. E. Bloom, S. K. McConnell, J. L. Roberts, N. C. Spitzer and M. J. Zigmond, eds., *Fundamental Neuroscience*, 2nd edn. Amsterdam: Academic Press, 533–554.

Woolfolk, R. L. 1999. Malfunction and mental disorder. *The Monist* 82: 658–670.

World Health Organization. 2001. *International Classification of Functioning, Disability and Health*. Geneva: World Health Organization Press.

Wouters, A. G. 1995. Viability explanation. *Biology and Philosophy* 10: 435–457.

2013. Biology's functional perspective: Roles, advantage, and organization. In K. Kampourakis, ed., *The Philosophy of Biology: A Companion for Educators.* Dordrecht: Springer. 455–486.

Wright, L. 1973. Functions. *Philosophical Review* 82: 139–168.

1976. *Teleological Explanations.* Berkeley: University of California Press.

2013. Epilogue. In P. Huneman, ed., *Function: Selection and Mechanisms.* 233–243. Dordrecht: Springer.

Young, J. Z. 1964. *A Model of the Brain.* Oxford: Clarendon Press.

Index

aboutness. *See* intentionality
Abrams, M., 191
Amundson, R., 141, 145–146, 148, 214
antibody selection, 69–72, 132
Ariew, A., 44, 168

backwards causation, 16, 48
Barros, B., 168
Bechtel, W., 162, 167
Bedau, M., 42, 49
biological information, 190
Blackman, R., 92, 213
Boorse, C., 46–47, 49, 134–138, 146–147
Bouchard, F., 69, 91, 95, 145, 150
Brandon, R., 130
Buller, D., 20, 57–59, 116, 176
Burge, T., 210–211

Campbell, D., 68, 74
Cao, R., 199
Caro, T., 9, 12–14, 20, 23, 46
Changeux, J.-P., 82, 87
conceptual analysis, 22–23
constancy mechanisms, 210–211
Craver, C., 45, 110, 122, 154, 160–161
Crick, F., 96
cultural evolution, 31
Cummins, R., 28, 32, 110, 122, 126, 146–147, 163

Daphnia, 177, 185, 190, 204
Darden, L., 69, 153–154, 165, 167
Darwin, C., 9, 66
Davies, P., 36–37, 126
Dennett, D., 76, 209
Developmental Origins of Health and Disease (DOHaD), 173
Dickinson, A., 188
Doolittle, W. F., 69, 95–96, 105, 143, 214
Dretske, F., 115, 185, 196, 206–211
dysfunction, 15–16, 28–29, 36–37, 94, 175

and biostatistical theory, 134–138
and causal role theory, 126
and functional indeterminacy, 111–114
and line drawing, 133
and mechanisms, 156
and normal environment, 129–132
and pandemic disease, 135
and value judgments, 124–125
vs. abnormal circumstances, 126–128

Eddy, S., 13
Enç, B., 202
ENCODE, 2–3, 143–144, 214
etiology, 15
evolutionary psychology, 158, 174, 176
extended cognition, 189
extended phenotype, 21

Fodor, J., 187, 201
Foucault, M., 186
function, biological. *See also* dysfunction
    and accidents, 10, 15, 28–29, 40–41, 90, 94, 164
    vs. artifacts, 21–22, 30–31
    and artifacts, 21–22, 30–31, 213
    and cancer cells, 52–53
    and clay crystals, 4, 42
    clay crystals, 4
    and design, 18–19
    direct and derived, 72–73, 91, 100–101, 196–197
    as epiphenomenal, 37
    response functions, 90, 175, 193
    and situation-specificity, 137
    vs. structure, 141–142
    and swamp creatures, 41
    and vestiges, 38–39
function, theories of
    biostatistical theory, 17, 46–47, 110, 132, 146–147
    causal role theory, 17, 110, 120, 126, 144–148

233

Printed in the United States
by Baker & Taylor Publisher Services